Dual Antiplatelet Therapy for Coronary and Peripheral Arterial Disease

Dual Antiplatelet Therapy for Coronary and Peripheral Arterial Disease

Edited by

Sorin J. Brener, MD

Professor of Medicine - Weill Cornell Medical College
Director, Cardiac Catheterization Laboratory
New York-Presbyterian Brooklyn Methodist Hospital
Division of Cardiology
New York, USA

Associate editor

Michael I. Brener, MD

Fellow, Cardiology, Columbia University
Medical Center-NewYork Presbyterian Hospital
New York, NY, United States

ELSEVIER

ACADEMIC PRESS
An imprint of Elsevier

Academic Press is an imprint of Elsevier
125 London Wall, London EC2Y 5AS, United Kingdom
525 B Street, Suite 1650, San Diego, CA 92101, United States
50 Hampshire Street, 5th Floor, Cambridge, MA 02139, United States
The Boulevard, Langford Lane, Kidlington, Oxford OX5 1GB, United Kingdom

Notices
Knowledge and best practice in this field are constantly changing. As new research and
experience broaden our understanding, changes in research methods, professional
practices, or medical treatment may become necessary.

Practitioners and researchers must always rely on their own experience and knowledge in
evaluating and using any information, methods, compounds, or experiments described
herein. In using such information or methods they should be mindful of their own safety
and the safety of others, including parties for whom they have a professional
responsibility.

To the fullest extent of the law, neither the Publisher nor the authors, contributors, or
editors, assume any liability for any injury and/or damage to persons or property as a
matter of products liability, negligence or otherwise, or from any use or operation of any
methods, products, instructions, or ideas contained in the material herein.

Library of Congress Cataloging-in-Publication Data
A catalog record for this book is available from the Library of Congress

British Library Cataloguing-in-Publication Data
A catalogue record for this book is available from the British Library

ISBN: 978-0-12-820536-5

For information on all Academic Press publications visit our website at
https://www.elsevier.com/books-and-journals

Publisher: Stacy Masucci
Acquisitions Editor: Katie Chan
Editorial Project Manager: Billie Jean Fernandez
Production Project Manager: Sreejith Viswanathan
Cover Designer: Matthew Limbert

Typeset by TNQ Technologies

Contents

Contributors

Dominick J. Angiolillo, MD, PhD
Professor of Medicine, Division of Cardiology, University of Florida College of Medicine, Jacksonville, FL, United States

Michael I. Brener, MD
Fellow, Cardiology, Columbia University Medical Center-NewYork Presbyterian Hospital, New York, NY, United States

Sorin J. Brener, MD
Professor of Medicine, NewYork Presbyterian-Brooklyn Methodist Hospital, Brooklyn, NY, United States

Davide Capodanno, MD, PhD
Professor of Cardiovascular Diseases, Division of Cardiology, A.O.U. Policlinico "G. Rodolico - San Marco", University of Catania, Catania, Italy

Bimmer Claessen, MD, PhD
Staff Physician, Center for Interventional Cardiovascular Research and Clinical Trials, Zena and Michael A. Wiener Cardiovascular Institute, Icahn School of Medicine at Mount Sinai, New York, NY, United States

Francesco Costa, MD, PhD
Staff Physician, Department of Clinical and Experimental Medicine, Policlinic "G. Martino", University of Messina, Messina, Italy

C. Michael Gibson, MD
Professor of Medicine, Division of Cardiovascular Medicine, Department of Medicine, Beth Israel Deaconess Medical Center, Harvard Medical School, Boston, MA, United States; Baim Institute for Clinical Research, Boston, MA, United States

Ridhima Goel, MD
Research Fellow, Center for Interventional Cardiovascular Research and Clinical Trials, Zena and Michael A. Wiener Cardiovascular Institute, Icahn School of Medicine at Mount Sinai, New York, NY, United States

Antonio Greco, MD
Staff Physician, Division of Cardiology, A.O.U. Policlinico "G. Rodolico - San Marco", University of Catania, Catania, Italy

Chang Hoon Lee, MD, PhD
Staff Physician, Division of Cardiology, University of Florida College of Medicine, Jacksonville, FL, United States

Roxana Mehran, MD
Professor of Medicine, Center for Interventional Cardiovascular Research and Clinical Trials, Zena and Michael A. Wiener Cardiovascular Institute, Icahn School of Medicine at Mount Sinai, New York, NY, United States

Nino Mihatov, MD
Research Fellow, Richard A. and Susan F. Smith Center for Outcomes Research, Beth Israel Deaconess Medical Center & Harvard Medical School, Boston, MA, United States; Clinical & Research Fellow in Cardiovascular Medicine, Division of Cardiology, Department of Medicine, Massachusetts General Hospital & Harvard Medical School, Boston, MA, United States

Johny Nicolas, MD
Research Fellow, Center for Interventional Cardiovascular Research and Clinical Trials, Zena and Michael A. Wiener Cardiovascular Institute, Icahn School of Medicine at Mount Sinai, New York, NY, United States

Adam T. Phillips, MD
Staff Physician, Division of Cardiovascular Medicine, Department of Medicine, Beth Israel Deaconess Medical Center, Harvard Medical School, Boston, MA, United States; Baim Institute for Clinical Research, Boston, MA, United States

Lauren S. Ranard, MD
Fellow, Cardiology, Columbia University Medical Center-NewYork Presbyterian Hospital, New York, NY, United States

Sudhakar Sattur, MD, MHSA
Staff Physician, Guthrie Clinic and Robert Packer Hospital, Sayre, PA, United States

Robert F. Storey, MD
Professor of Medicine, Department of Infection, Immunity and Cardiovascular Disease, University of Sheffield, Sheffield, United Kingdom

Marco Valgimigli, MD, PhD
Professor of Medicine, Swiss Cardiovascular Center Bern, Bern University Hospital, Bern, Switzerland

Freek W.A. Verheugt, MD, PhD
Professor of Medicine, Department of Cardiology, Heart Center, Onze Lieve Vrouwe Gasthuis (OLVG), Amsterdam, Netherlands

Robert Yeh, MD, MSc
Associate Professor of Medicine, Division of Cardiology, Department of Medicine, Beth Israel Deaconess Medical Center, Harvard Medical School, Boston, MA, United States

Preface

Dear colleague,

On behalf of Elsevier and the Editors, I welcome you to this compendium of DUAL ANTIPLATELET THERAPY FOR CORONARY AND PERIPHERAL ARTERIAL DISEASE. We attempted to create a new paradigm of rapid publication of updated information in an ever-evolving field. Albeit delayed somewhat by the COVID-19 pandemic, we were still able to produce in a few short months an up-to-date collection of monographs from world experts in vascular disease from Italy, Switzerland, the Netherlands, the United Kingdom, and the United States.

After an extensive review of the pathophysiology of atherosclerosis and its manifestations in various vascular beds, Angiolillo and colleagues discuss relevant aspects of platelet physiology and pathophysiology in order to set the stage for the utilization of dual antiplatelet therapy (DAPT) in clinical practice. Phillips and Gibson complete this discussion with the history of DAPT development and the various mechanisms of action that are pertinent to clinical practice.

In Chapter 4, Lauren Ranard updates us on the most common utilization of DAPT—patients with acute coronary syndromes, while Sudhakar Sattur, and Antonio Greco and Davide Capodanno take on the less common fields for DAPT use, the peripheral and cerebral circulation, respectively.

In Chapters 7 and 8, Marco Valgimgli and Michael Brener tackle the thorny issue of DAPT duration in various clinical scenarios using recent randomized clinical trials and meta-analyses summarizing them. The difficult-to-achieve balance between prevention of ischemic events and causation of clinically relevant bleeding is addressed by Robert Yeh and colleagues in Chapter 9.

Roxana Mehran and her group address the novel concept of monotherapy instead of DAPT for patients undergoing PCI or presenting with ACS in Chapter 10.

The book is completed by Freek Verheught and Robert Storey who review the accumulating evidence on the combination of antiplatelet and antithrombotic agents in patients with indications for both. The conundrum of when "less is more" is cogently resolved in Chapter 11.

We hope you enjoy!

Platelets and arterial disease—initiation, progression, and destabilization of atherosclerotic vascular disease

Sorin J. Brener, MD

Professor of Medicine, NewYork Presbyterian-Brooklyn Methodist Hospital, Brooklyn, NY, United States

Atherosclerosis is responsible for the vast majority of ischemic events affecting the cerebral, cardiac, and peripheral circulation. Its main consequences, i.e., acute ischemic stroke, acute myocardial infarction, chronic angina, acute limb ischemia, and intermittent claudication, affect millions of individuals around the world, predominantly in the older age group. Decades of research in the basic sciences and years of clinical experience have elucidated various aspects of atherosclerotic cardiovascular disease (ASCVD), including its risk factors, molecular mechanisms, and potential targets for therapy. We now recognize that an incredibly complex mosaic of interactions between components of the vascular and immune systems modulates the action and location of constitutive blood components, resulting in thrombosis and compromised arterial flow to essential organs. A lipid-centric theory to explain the interactions underlying atherosclerosis[1] has been replaced by a thrombus-focused paradigm, only to be displaced by our current understanding of the process as a confluence of inflammation and thrombosis leading to the initiation of the atherosclerotic plaque and its subsequent progression and destabilization (Fig. 1.1).

This book examines in detail the various effects of platelets on the process of ASCVD and the potential benefits of antiplatelet therapy in the prevention of ischemic events. It is normal then to ask whether the platelets are involved themselves in the initiation of atherosclerosis or are merely reacting to its presence. Do they promote its progression once it started and can we arrest the process by inhibiting some of their functions? To address some of these questions, this chapter will focus on three phases of atherosclerosis, namely, initiation of the disease process, transition to vulnerable lesions, and thrombosis of the ruptured plaque, before summarizing the role platelets play in these intricate mechanisms.

Dual Antiplatelet Therapy for Coronary and Peripheral Arterial Disease
https://doi.org/10.1016/B978-0-12-820536-5.00009-4

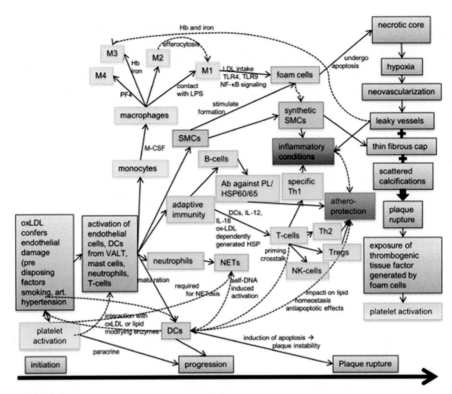

FIGURE 1.1

Overview of atherosclerosis. *Ab*, antibody; *DCs*, dendritic cells; *Hb*, hemoglobin; *HSP*, heat shock protein; *IL*, interleukin; *LDL*, low-density lipoprotein; *M-CSF*, macrophage colony-stimulating factor; *NET*, neutrophil extracellular trap; *NF-κB*, nuclear factor κB; *NK*, natural killer; *oxLDL*, oxidized low-density lipoprotein; *PF4*, platelet factor 4; *SMC*, smooth muscle cell; *Treg*, regulatory T cell; *VALT*, vascular-associated lymphatic tissue.

Reproduced with permission from Langer et al. in Front Immunol. *2015, article 98.*

Initiation of atherosclerotic disease

ASCVD develops in all arterial beds but most of our understanding of the process emerged from in vivo and in vitro studies of coronary arteries. ASCVD develops in the presence of endothelial dysfunction and circulating low-density lipoprotein (LDL) cholesterol particles; each factor is necessary but not sufficient. The factors promoting endothelial dysfunction are the classical risk factors identified as precursors of coronary artery disease, namely, systemic arterial hypertension, hyperlipidemia, diabetes mellitus, smoking, and genetic predisposition. The damaged endothelium changes its permeability and allows LDL particles to enter the subintimal space, where they undergo aggregation, fusion, oxidation, and incorporation into complexes that promote atherogenesis.[2] These modified LDL particles cause the endothelium to secrete chemokines and selectins that recruit leukocytes (mononuclear cells and lymphocytes) and facilitate their migration into the subendothelial space, where they express scavenger receptors and transform first into macrophages and then later into foam cells after ingesting LDL (Fig. 1.2).

Platelets play an important role in this process and mediate its pace and extent. Normally, three pathways provide constant protection from platelet adhesion to endothelium: nitric oxide (NO), ecto-ADPase, and prostaglandin I$_2$.[3] It is notable that just as activated platelets can bind to intact endothelium, restive platelets can also bind to activated endothelial cells. This adhesion to lesion-free endothelium at atherosclerosis-prone sites (such as bifurcation of coronary branches or areas with nonlaminar flow pattern) is mediated via P-selectin. Even brief tethering engenders platelet activation and further binding.[4] Platelet adhesion is significant because it leads to secretion of signal molecules and the migration and extravasation of inflammatory cells to that site. As soon as endothelial damage from atherosclerosis risk factors occurs, the exposure of collagen in the subendothelial matrix tethers platelets to the vascular surface via an interaction between von Willebrand factor (vWF) secreted from the endothelium and the glycoprotein (GP) Ib-factor V-factor IX complex on the platelet surface. The platelets then roll (mediated by P-selectin) on the endothelial surface and are anchored via the GP VI complex. This connection modifies platelets' shape and activates them, leading to degranulation and secretion of thromboxane (TX) and adenosine diphosphate (ADP). Activated platelets ingest circulating LDL and these "fat" platelets are immediately engulfed

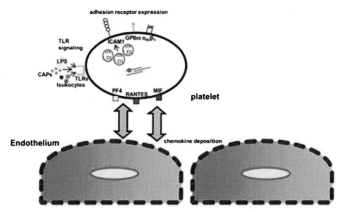

FIGURE 1.2

Initiation of atherosclerosis. *GP*, glycoprotein; *ICAM1*, intracellular adhesion molecule 1; *LPS*, lipopolysaccharide; *MIF*, macrophage migratory inhibitory factor; *oxLDL*, oxidized low-density lipoprotein; *PF4*, platelet factor 4; *RANTES*, chemokine ligand 5 (CCL5); *TLRs*, toll-like receptors.

Reproduced with permission from Langer et al. in Front Immunol. *2015, article 98.*

by foam cells and are further activated by LDL. This is a critical step in the development of a lipid-rich plaque.[5]

The most important aspect of platelets' contribution to ASCVD is their interaction with inflammatory cells. When activated, platelets secrete proteins and compounds from two important types of granules: (1) α granules, which contain platelet factor 4 (PF4), factor V, platelet-derived growth factor, and epidermal growth factor and are critical to platelet aggregation, and (2) dense granules, which contain mostly calcium, ADP, and adenosine triphosphate and are important for the amplification of the prothrombotic signal.[6] Tethered and anchored platelets deliver PF4 and "regulated on activation, normal T-cell expressed, and presumably secreted" (RANTES or CC5) to the endothelial surface and monocytes, respectively, leading to macrophage infiltration of the vascular wall.[7] PF4 induces transformation of monocytes into macrophages. While RANTES regulates the adhesion of monocytes, this interaction is modulated by P-selectin, a platelet-derived compound. Activated platelets and endothelial cells secrete other integrins, such CD 40L and interleukin 1β, causing activation of nuclear factor κB and a plethora of interactions resulting in monocyte adhesion and transmigration.[8] Activated platelets trapped in the atherosclerotic plaque provide a continuous supply of interleukin 1β, perpetuating inflammation.[9] Moreover, platelets' direct interaction with monocytes leads to firmer adhesion of monocytes to the damaged endothelium.[10] The adaptive immune system, T lymphocytes, in particular, plays a role in platelet activation, as does the innate immune system, where complement factors are thought to promote binding of inflammatory cells to the endothelium.[11,12]

Other cellular components, in addition to platelets and inflammatory cells, have been identified in the vicinity of the damaged endothelium. Progenitor cells derived from the bone marrow adhere to platelets, expressing GP IIb/IIIa receptors and P-selectin, and can influence vascular repair or remodeling. In turn, the activated platelets secrete stromal cell-derived factor 1α (SDF-1α) that enhances the binding of progenitor cells to a developing arterial thrombus.[4,13] Statins and thiazolidinediones can prevent this transformation of progenitor cells into foam cells, thus arresting the atherosclerotic process.[14]

Besides the cellular mechanisms described earlier, the rheology of blood flow is critically important. Local blood flow conditions significantly affect the interaction of platelets with endothelial cells and with leukocytes, particularly at atherosclerosis-prone sites, such as bifurcations. While most of the flow in coronary arteries is laminar, the different fluid planes travel at various speeds because of fluid drag forces exerted by the vessel wall. When they reach bifurcation points, a host of flow disturbances ensue and create areas with low shear stress (<20 dynes/cm^2) or high shear stress (>2000 dynes/cm^2). At very high shear stress (>5000 dynes/cm^2), as seen in narrowed or very tortuous arteries, platelet activation is induced without additional triggers,[15] which promotes secretion of vWF and ADP[16] and is not inhibited by cyclooxygenase and ADP pathways.[17] In contrast, low shear stress (~ 150 dynes/cm^2) is more conducive for leukocyte adhesion to the endothelium, contributing to atherosclerosis initiaition.[18]

Transition to vulnerable plaque

After decades of plaque deposition (Fig. 1.3) in arterial walls, described earlier, some patients develop symptoms related to the sudden destabilization of these deposits. The processes responsible for this transition and their manifestations are myriad and beyond the scope of this chapter. We will concentrate the following remarks on the role of platelets in this transformation of stable, lipid-laden plaques into the classical atherothrombotic lesion responsible for most acute myocardial infarctions. To understand how platelets may contribute to this process, it is worth reviewing the features of the "vulnerable plaque" phenotype—a collection of parameters gleaned from autopsy studies, atherectomy coronary specimens, intravascular ultrasound and optical coherence tomographic imaging, and specimens from operated carotid arteries.[19–22] The principal characteristic is the lipid-rich core, a conglomerate of oxidized lipid, rich in tissue factor (TF), which became devoid of cells after apoptosis of vascular smooth muscle cells (VSMCs) and macrophages.[23] A fibrous cap protects this amorphous material from contact with the blood stream. It is composed of inflammatory cells (macrophages) and VSMCs, which secrete extracellular matrix. The heaviest inflammatory cell concentration is at the shoulders of the plaque. In contrast, the core of the plaque is rather hypocellular, with little inflammation.[24] The cap is thinned by collagen lysis caused by matrix metalloproteinases (MMPs) secreted from these inflammatory cells.[25] Activated platelets secrete MMP-2 while aggregating,[26] and the binding of platelets to activated endothelium leads to endothelial synthesis and secretion of MMP-9, a direct contributor to fibrous cap thinning.[27] Besides the frequent calcifications in the intima and tunica media of atherosclerotic vessels, the plaques are distinguished from other tissue by a network of vasa vasorum supplying blood to the growing atheroma, some of it incited by plaque hypoxia.[28] All these components coexist in the setting of an active positive remodeling process leading to expansion of the vessel cross-sectional area to limit the impact of the growing plaque.[29] Thus, the principal contributions of platelets to the destabilization of the plaque are the recruitment of inflammatory cells in the fibrous cap and the production of MMPs that disturb the balance of collagen formation and degradation in favor of the latter (Fig. 1.4).

Thrombosis of the ruptured plaque

Once the atherosclerotic plaque became unstable, an intricate cascade of events and interactions lead to the formation of a thrombus (Fig. 1.5). The fate of this fresh thrombus is dictated by the balance between local thrombotic and fibrinolytic stimuli. If the former predominates, the thrombus will grow and eventually occlude the artery with ensuing ischemia and infarction. If the latter prevails, the thrombus may dissipate or decrease in size and potentially participate in the

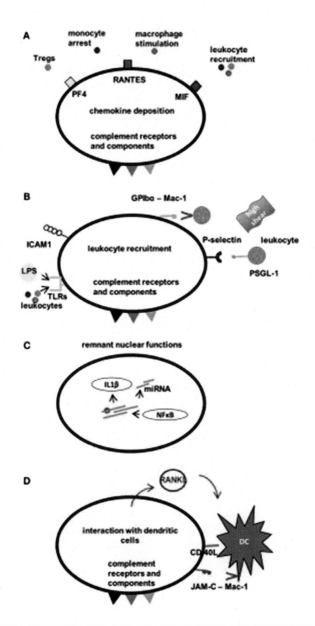

FIGURE 1.3

Progression of atherosclerosis. Platelets contribute to the progression of atherosclerosis by (A) chemokine deposition and (B) leukocyte recruitment. Remnant "nuclear" functions of platelets, such as the ability to translate and modify messenger RNA, promote inflammation and endothelial polarization (C). Furthermore, the interaction with antigen-presenting dendritic cells can contribute to atheroprogression involving adhesion receptors and soluble mediators (D). *DC*, dendritic cell; *GP*, glycoprotein; *ICAM1*, intracellular adhesion molecule 1; *IL*, interleukin; *LPS*, lipopolysaccharide; *miRNA*, microRNA; *MIF*, macrophage migratory inhibitory factor; *NF-κB*, nuclear factor κB; *PF4*, platelet factor 4; *RANTES*, chemokine ligand 5 (CCL5); *TLRs*, toll-like receptors; *Treg*, regulatory T cell.

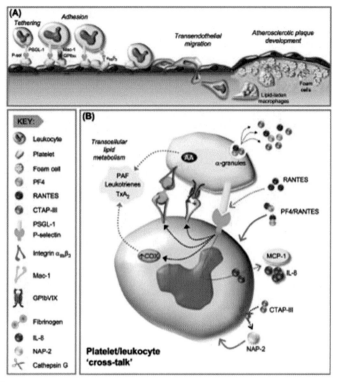

FIGURE 1.4

Platelet-leukocyte interaction. (A) Platelet-mediated leukocyte recruitment and adhesion.
(B) Platelet-leukocyte cross talk. *IL*, interleukin; *PAF*, platelet activating factor; *PF4*,
platelet factor 4; *RANTES*, chemokine ligand 5 (CCL5); TXA_2, thromboxane A_2.

Reproduced with permission from Jackson et al. Hematology 2011:61–61.

healing process of the injured segment and the stepwise increase in arterial stenosis. It is estimated that many more plaque rupture events result in fibrinolysis
and healing than in thrombosis and infarction. How do platelets participate in this
elaborate process?

Just like during atherosclerosis formation, platelets are attracted to the
injured endothelium after plaque rupture and interact with subendothelial
collagen and vWF to adhere and become activated, which leads to degranulation.
The initial contact is made between vWF array A1 and GP 1bα,[30] leading to tethering and binding to collagen, fibronectin, and laminin.[31] Beyond the release of
thrombin after degranulation, other compounds are also exteriorized. Large
amounts of P-selectin are released from the platelets together with TXA_2 and
the platelet membrane undergoes structural changes. TXA_2 binds to its widely
distributed receptor on platelets, endothelial cells, and inflammatory cells,
unleashing a powerful amplification reaction, such that more platelets become

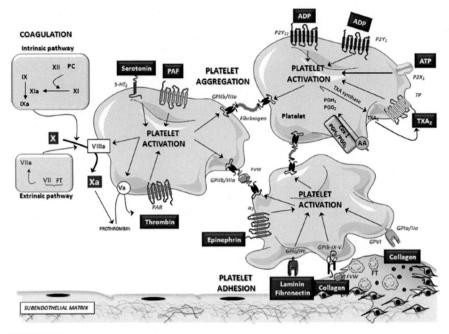

FIGURE 1.5

Overview of the mechanism of ruptured plaque thrombosis. *AA*, arachidonic acid; *ADP*, adenosine diphosphate; *ATP*, adenosine triphosphate; *GP*, glycoprotein; *PAF*, platelet activating factor; *PAR*, protease-activated receptor; *PC*, protein C; *PG*, prostaglandin; *TF*, tissue factor; *TP*, thromboxane receptors; *TXA_2*, thromboxane A$_2$; *vWF*, von Willebrand factor.

Reproduced with permission from Badimon et al. Eur Heart J: Acute Cardiovasc Care. 2012;1:60–74.

activated and bind to each other via fibrinogen molecules attaching to the GP IIb/IIIa. In parallel, the soluble coagulation cascade is activated by TF released from exposed foam cells. It binds circulating factors VII/VIIa and activates the extrinsic pathway of coagulation (Fig. 1.6).[32] Platelets accentuate this process by expressing phospholipids on their surface, which serve as "cooktop" for coagulation factors to become activated and bind to each other. Besides phospholipids, platelets synthetize proteins in response to activation and aggregation. Among these ∼300 products are the proinflammatory interleukin 1β and the prothrombotic TF.[33,34] The injured endothelium completes the process by switching to a procoagulant phenotype and secreting tissue plasminogen activator inhibitor 1 (PAI-1) (Fig. 1.7).

From an epidemiologic perspective, patients with risk factors for atherosclerosis have been considered to also have an increased thrombotic risk. This is manifested by hypercoagulability, hypofibrinolysis, or exceedingly reactive platelets. The last risk factor is particularly relevant to diabetic patients in whom platelet reactivity

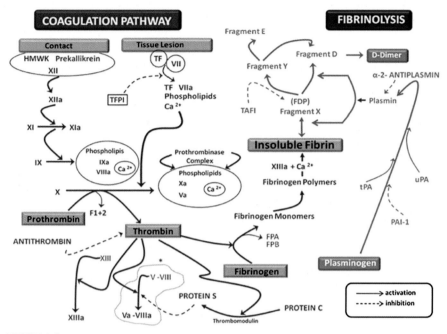

FIGURE 1.6

Contribution of the soluble coagulation cascade to ruptured plaque thrombosis. *FPA*, fibrinopeptide A; *FPB*, fibrinopeptide B; *FDP*, fibrin degradation products; *HMWK*, high-molecular-weight kininogen; *PAI-1*, plasminogen activator inhibitor type 1; *TF*, tissue factor; *TFPI*, tissue factor pathway inhibitor; *tPA*, tissue plasminogen activator; *uPA*, urokinase plasminogen activator.

Reproduced with permission from Badimon et al. Eur Heart J: Acute Cardiovasc Care. *2012;1:60–74.*

has been directly linked to lack of optimal blood glucose control. Confirming this relationship is the observation that improvement in risk factors diminishes platelet reactivity.[35,36] It is possible that much of the increased platelet reactivity in diabetes is mediated by heightened levels of endothelial cells apoptosis, which activates platelets and monocytes.[37]

Another set of important observations concluded that platelets may promote both atherosclerosis and thrombosis via platelet microparticles (PMps). These are secreted locally from activated platelets as microvesicles.[38] PMps are pro-coagulant by virtue of their ability to expose anionic phospholipids on the platelet membrane, which are essential for the assembly of the coagulation factors. Furthermore, PMps may inhibit fibrinolysis by enhancing PAI-1 activity.[39] PMps harbor a large array of chemokines and adhesion receptors that mediate interaction between platelets and inflammatory cells, as well as between platelet and endothelial cells. They likely are the vehicle for transferring RANTES to the endothelium afflicted by atherosclerosis.[40]

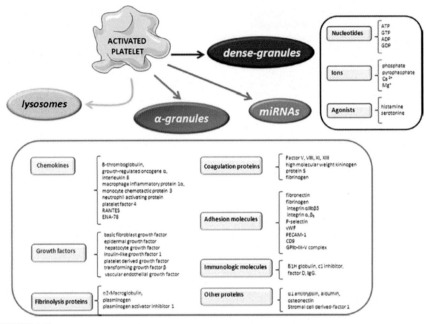

FIGURE 1.7

Molecules released from activated platelets. *ADP*, adenosine diphosphate; *ATP*, adenosine triphosphate; *ENA-78*, epithelial cell-derived neutrophil-activating peptide 78; *GDT*, guanosine diphosphate; *GTP*, guanosine triphosphate; *miRNA*, microRNA; *PECAM-1*, platelet endothelial cellular adhesion molecule-type 1; *RANTES*, regulated on activation, normal T-cell expressed, and presumably secreted; *vWF*, von Willebrand factor.

Reproduced with permission from Badimon et al. Eur Heart J: Acute Cardiovasc Care. 2012;1:60–74.

Summary

Platelets are important contributors to the process of atherothrombosis, both at its initial stage and in its most advanced and dramatic form, i.e., acute coronary thrombosis. Intensive research into the role of platelets has identified them as an intricate source of substances that promote adhesion to intact and activated endothelium and facilitate the binding of inflammatory cells to atherosclerosis-prone sites. The latter is probably the most important contribution of platelets to this process. This wealth of evidence informs our attempts to target specific aspects of platelet function in order to decrease the incidence of ischemic events. Traditional approaches such as inhibition of cyclooxygenase, ADP, or GP IIb/IIIa pathways have proved insufficient in this endeavor, opening the door to more novel and daring modifications of these complex interactions. For example, platelet adhesion to endothelium has been blocked in vitro by antibodies against vWF.[41,42] Inhibition of the SDF-1α receptor

(CXCR4) prevented platelet aggregation and mitigated thrombosis on activated endothelium.[43] Attempts to modify the function of platelet-derived RANTES with a chemically engineered variant resulted in reduced infiltration with inflammatory cells and lower production of MMP-9 in a mice LDL receptor knockout model. In contrast, there was a significant increase in smooth muscle cells and collagen levels, resulting in a more stable plaque.[44] These and other approaches are likely to take better advantage of our increasing understanding of atherothrombosis biology and the pivotal role platelets play in its initiation and progression.

References

1. Konstantinov IE, Mejevoi N, Anichkov NM. Nikolai N. Anichkov and his theory of atherosclerosis. *Tex Heart Inst J*. 2006;33(4):417−423.
2. Llorente-Cortes V, Badimon L. LDL receptor-related protein and the vascular wall: implications for atherothrombosis. *Arterioscler Thromb Vasc Biol*. 2005;25(3):497−504.
3. Jin RC, Voetsch B, Loscalzo J. Endogenous mechanisms of inhibition of platelet function. *Microcirculation*. 2005;12(3):247−258.
4. Massberg S, Brand K, Gruner S, et al. A critical role of platelet adhesion in the initiation of atherosclerotic lesion formation. *J Exp Med*. 2002;196(7):887−896.
5. Daub K, Langer H, Seizer P, et al. Platelets induce differentiation of human CD34+ progenitor cells into foam cells and endothelial cells. *FASEB J*. 2006;20(14):2559−2561.
6. Palomo I, Toro C, Alarcon M. The role of platelets in the pathophysiology of atherosclerosis (review). *Mol Med Rep*. 2008;1(2):179−184.
7. Pitsilos S, Hunt J, Mohler ER, et al. Platelet factor 4 localization in carotid atherosclerotic plaques: correlation with clinical parameters. *Thromb Haemostasis*. 2003;90(6):1112−1120.
8. Bavendiek U, Libby P, Kilbride M, Reynolds R, Mackman N, Schonbeck U. Induction of tissue factor expression in human endothelial cells by CD40 ligand is mediated via activator protein 1, nuclear factor ka B, and Egr-1. *J Biol Chem*. 2002;277(28):25032−25039.
9. Lindemann S, Tolley ND, Dixon DA, et al. Activated platelets mediate inflammatory signaling by regulated interleukin 1beta synthesis. *J Cell Biol*. 2001;154(3):485−490.
10. Weber C, Springer TA. Neutrophil accumulation on activated, surface-adherent platelets in flow is mediated by interaction of Mac-1 with fibrinogen bound to alphaIIbbeta3 and stimulated by platelet-activating factor. *J Clin Invest*. 1997;100(8):2085−2093.
11. Manthey HD, Zernecke A. Dendritic cells in atherosclerosis: functions in immune regulation and beyond. *Thromb Haemostasis*. 2011;106(5):772−778.
12. Hamad OA, Nilsson PH, Wouters D, Lambris JD, Ekdahl KN, Nilsson B. Complement component C3 binds to activated normal platelets without preceding proteolytic activation and promotes binding to complement receptor 1. *J Immunol*. 2010;184(5):2686−2692.
13. Massberg S, Konrad I, Schurzinger K, et al. Platelets secrete stromal cell-derived factor 1alpha and recruit bone marrow-derived progenitor cells to arterial thrombi in vivo. *J Exp Med*. 2006;203(5):1221−1233.

14. Wang C-H, Ciliberti N, Li S-H, et al. Rosiglitazone facilitates angiogenic progenitor cell differentiation toward endothelial lineage. *Circulation*. 2004;109(11):1392−1400.

15. Savage B, Saldivar E, Ruggeri ZM. Initiation of platelet adhesion by arrest onto fibrinogen or translocation on von Willebrand factor. *Cell*. 1996;84(2):289−297.

16. Goto S, Tamura N, Handa S. Effects of adenosine 5'-diphosphate (ADP) receptor blockade on platelet aggregation under flow. *Blood*. 2002;99(12):4644−4645.

17. Nesbitt WS, Westein E, Tovar-Lopez FJ, et al. A shear gradient-dependent platelet aggregation mechanism drives thrombus formation. *Nat Med*. 2009;15(6):665−673.

18. Woollard KJ, Kling D, Kulkarni S, Dart AM, Jackson S, Chin-Dusting J. Raised plasma soluble P-selectin in peripheral arterial occlusive disease enhances leukocyte adhesion. *Circ Res*. 2006;98(1):149−156.

19. Virmani R, Kolodgie FD, Burke AP, Farb A, Schwartz SM. Lessons from sudden coronary death: a comprehensive morphological classification scheme for atherosclerotic lesions. *Arterioscler Thromb Vasc Biol*. 2000;20(5):1262−1275.

20. Moreno PR, Falk E, Palacios IF, Newell JB, Fuster V, Fallon JT. Macrophage infiltration in acute coronary syndromes. Implications for plaque rupture. *Circulation*. 1994;90(2): 775−778.

21. Kubo T, Imanishi T, Takarada S, et al. Assessment of culprit lesion morphology in acute myocardial infarction: ability of optical coherence tomography compared with intravascular ultrasound and coronary angioscopy. *J Am Coll Cardiol*. 2007;50(10):933−939.

22. Spagnoli LG, Mauriello A, Sangiorgi G, et al. Extracranial thrombotically active carotid plaque as a risk factor for ischemic stroke. *J Am Med Assoc*. 2004;292(15):1845−1852.

23. Felton CV, Crook D, Davies MJ, Oliver MF. Relation of plaque lipid composition and morphology to the stability of human aortic plaques. *Arterioscler Thromb Vasc Biol*. 1997;17(7):1337−1345.

24. Schwartz SM, Galis ZS, Rosenfeld ME, Falk E. Plaque rupture in humans and mice. *Arterioscler Thromb Vasc Biol*. 2007;27(4):705−713.

25. Galis ZS, Sukhova GK, Lark MW, Libby P. Increased expression of matrix metalloproteinases and matrix degrading activity in vulnerable regions of human atherosclerotic plaques. *J Clin Invest*. 1994;94(6):2493−2503.

26. Fernandez-Patron C, Martinez-Cuesta MA, Salas E, et al. Differential regulation of platelet aggregation by matrix metalloproteinases-9 and -2. *Thromb Haemostasis*. 1999;82(6):1730−1735.

27. May AE, Kalsch T, Massberg S, Herouy Y, Schmidt R, Gawaz M. Engagement of glycoprotein IIb/IIIa (alpha(IIb)beta3) on platelets upregulates CD40L and triggers CD40L-dependent matrix degradation by endothelial cells. *Circulation*. 2002;106(16): 2111−2117.

28. Slevin M, Turu MM, Rovira N, et al. Identification of a 'snapshot' of co-expressed angiogenic markers in laser-dissected vessels from unstable carotid plaques with targeted arrays. *J Vasc Res*. 2010;47(4):323−335.

29. Schoenhagen P, Ziada KM, Kapadia SR, Crowe TD, Nissen SE, Tuzcu EM. Extent and direction of arterial remodeling in stable versus unstable coronary syndromes : an intravascular ultrasound study. *Circulation*. 2000;101(6):598−603.

30. Ruggeri ZM. Structure and function of von Willebrand factor. *Thromb Haemostasis*. 1999;82(2):576−584.

31. Nieswandt B, Watson SP. Platelet-collagen interaction: is GPVI the central receptor? *Blood*. 2003;102(2):449−461.

32. Toschi V, Gallo R, Lettino M, et al. Tissue factor modulates the thrombogenicity of human atherosclerotic plaques. *Circulation*. 1997;95(3):594−599.
33. Denis MM, Tolley ND, Bunting M, et al. Escaping the nuclear confines: signal-dependent pre-mRNA splicing in anucleate platelets. *Cell*. 2005;122(3):379−391.
34. Schwertz H, Tolley ND, Foulks JM, et al. Signal-dependent splicing of tissue factor pre-mRNA modulates the thrombogenicity of human platelets. *J Exp Med*. 2006;203(11): 2433−2440.
35. Osende JI, Badimon JJ, Fuster V, et al. Blood thrombogenicity in type 2 diabetes mellitus patients is associated with glycemic control. *J Am Coll Cardiol*. 2001;38(5):1307−1312.
36. Vivas D, Garcia-Rubira JC, Bernardo E, et al. Effects of intensive glucose control on platelet reactivity in patients with acute coronary syndromes. Results of the CHIPS Study ("Control de Hiperglucemia y Actividad Plaquetaria en Pacientes con Syndrome Coronario Agudo"). *Heart*. 2011;97(10):803−809.
37. Koga H, Sugiyama S, Kugiyama K, et al. Elevated levels of remnant lipoproteins are associated with plasma platelet microparticles in patients with type-2 diabetes mellitus without obstructive coronary artery disease. *Eur Heart J*. 2006;27(7):817−823.
38. Muller I, Klocke A, Alex M, et al. Intravascular tissue factor initiates coagulation via circulating microvesicles and platelets. *FASEB J*. 2003;17(3):476−478.
39. Forlow SB, McEver RP, Nollert MU. Leukocyte-leukocyte interactions mediated by platelet microparticles under flow. *Blood*. 2000;95(4):1317−1323.
40. Mause SF, von Hundelshausen P, Zernecke A, Koenen RR, Weber C. Platelet microparticles: a transcellular delivery system for RANTES promoting monocyte recruitment on endothelium. *Arterioscler Thromb Vasc Biol*. 2005;25(7):1512−1518.
41. Etingin OR, Silverstein RL, Hajjar DP. von Willebrand factor mediates platelet adhesion to virally infected endothelial cells. *Proc Natl Acad Sci USA*. 1993;90(11):5153−5156.
42. Gawaz M, Brand K, Dickfeld T, et al. Platelets induce alterations of chemotactic and adhesive properties of endothelial cells mediated through an interleukin-1-dependent mechanism. Implications for atherogenesis. *Atherosclerosis*. 2000;148(1):75−85.
43. Abi-Younes S, Sauty A, Mach F, Sukhova GK, Libby P, Luster AD. The stromal cell-derived factor-1 chemokine is a potent platelet agonist highly expressed in atherosclerotic plaques. *Circ Res*. 2000;86(2):131−138.
44. Braunersreuther V, Steffens S, Arnaud C, et al. A novel RANTES antagonist prevents progression of established atherosclerotic lesions in mice. *Arterioscler Thromb Vasc Biol*. 2008;28(6):1090−1096.

Platelet physiology and pharmacology—relevant considerations for patient care

Chang Hoon Lee, MD, PhD, Dominick J. Angiolillo, MD, PhD

Division of Cardiology, University of Florida College of Medicine, Jacksonville, FL, United States

Introduction

Platelets have an important role in the formation of hemostatic thrombi that develop at the site of vascular injury.[1] However, pathologic activation of platelets can lead to obstructive thrombi resulting in clinical manifestations such as acute myocardial infarction (MI), stroke, and acute limb ischemia. Therefore antiplatelet therapy has a key role in the treatment of cardiovascular (CV) disease manifestations. However, antiplatelet agents are associated with an increased risk of bleeding. Thus the ideal antiplatelet agent should have an optimal balance between safety and efficacy. Advancements in interventional vascular techniques have been paralleled by the development of antiplatelet treatment regimens, which often imply combining different therapies.[2] While most of these regimens have been developed in addition to aspirin therapy, which is considered the cornerstone of antiplatelet therapy for the treatment of CV disease manifestations, most recent studies have suggested that alternative regimens without aspirin (defined as "aspirin free") to be associated with a better safety profile while preserving efficacy.[3] This chapter provides an overview on platelet physiology and its implications for the development and use of antiplatelet therapies in patients with atherothrombotic disease manifestations.

Platelet physiology

Signaling processes of platelets involved in thrombus formation have been extensively investigated. We provide an overview of the structure and physiology of platelets and describe the mechanisms of activation that have set the basis for the development of antiplatelet agents.

Formation and structure

Pluripotent stem cells from the bone marrow differentiate into megakaryocytes. When the cytoplasm of a megakaryocyte is divided into proplatelets with specialized membranes, platelets are formed and released.[4] Platelets are the smallest blood cells (2−4 μm in diameter) and numbering $150-400 \times 10^9$/L in healthy individuals. Platelets have a short lifespan in the circulation (7−10 days) and are replaced at a rate of about 10% per day.[5]

In general, the structure of a platelet can be divided into four parts: peripheral zone, structural zone, organelle zone, and membranous zone.[6] The peripheral zone is the platelet membrane containing abundant glycoproteins (GPs) and phospholipids. The platelet membrane is rich in receptors, including adhesive receptors such as GP Ib-IX-V, GP VI, and integrin $\alpha_{IIb}\beta_3$ (known as GPIIb/IIIa) and activating receptors such as protease-activated receptor (PAR) 1, PAR4, $P2Y_{12}$, $P2Y_1$, and thromboxane prostanoid (TP) (Table 2.1, Fig. 2.1).[7−10] The structural zone contains microtubules, actin, and myosin, which maintain the shape of a platelet and have a role in facilitating alpha (α) and dense (δ) granule secretion through canalicular opening of the membranous zone during signaling activating process.[8] The organelle zone is composed of α-granules, δ-granules, and lysosomal (λ) granules. Alpha (α) granules contain adhesive membrane receptor, coagulation factors, von Willebrand factor (vWF), and fibrinogen. Dense (δ) granules contain high concentrations of adenine nucleotides (i.e., adenosine diphosphate (ADP)), cations (i.e., Ca^{2+}), and amines (i.e., histamine)). The granules and their contents are summarized in Table 2.2.

Table 2.1 Major platelet receptors, activators, and function.

Receptors	Activator	Function
Adhesive		
GP IIb/IIIa	Fibrinogen, vWF	Granule secretion, platelet spreading, clot retraction
GP1b-IX-V	vWF	TXA_2 synthesis, activating GPCR signaling, granule secretion
GP VI	Collagen	Ca^{2+} release, granule secretion, integrin activation
Activating		
PAR1	Thrombin (low level)	Ca^{2+} release, granule secretion, platelet shape change
PAR4	Thrombin (high level)	Same as PAR1 but less sensitive than PAR4, enhanced cleavage by PAR1, dimerizes with $P2Y_{12}$
$P2Y_1$	ADP	Ca^{2+} release, granule secretion
$P2Y_{12}$	ADP	Inhibition of adenylate cyclase (cAMP level decrease), granule secretion
TP	TXA_2	Ca^{2+} release, granule secretion, platelet shape change

ADP, *adenosine diphosphate*; cAMP, *cyclic adenosine monophosphate*; GP, *glycoprotein*; GPCR, *G protein-coupled receptor*; PAR, *protease-activated receptor*; TP, *thromboxane prostanoid*; TXA_2, *thromboxane A_2*; vWF, *von Willebrand factor*.

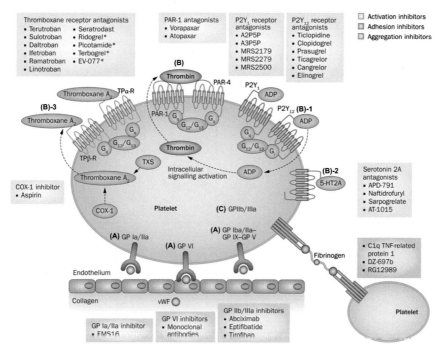

FIGURE 2.1 Platelet function and molecular targets of current and novel antiplatelet agents.

(A) Initiation phase. Platelet adhesion at the site of vascular injury is mediated by the binding of von Willebrand factor (vWF) to the platelet surface glycoprotein (GP) Ib-IX-V complex, which allows the binding of exposed collagen to GP VI and GP Ia/IIb. (B) Extension phase. Thrombin, generated on the surface of the activated platelet, is the most potent activator of human platelet through protease-activated receptor (PAR) 1 and PAR4. Three agonists from the activated platelet play important roles in positive feedback loops for amplifying the activation of platelet: (1) adenosine diphosphate (ADP) released form platelet dense granules triggers $P2Y_1$ and $P2Y_{12}$; (2) 5-hydroxytryptamine (5-HT; serotonin) released from platelet dense granules stimulates $5-HT_{2A}$ receptors; and (3) thromboxane A_2 generated by cyclooxygenase (COX)-1 stimulates thromboxane prostanoid (TP) receptor. (C) Perpetuation phase. Perpetuation of platelet-to-platelet aggregation is mediated by fibrinogen and vWF binding to activated integrin $\alpha_{IIb}\beta_3$ (GP IIb/IIIa). Current and emerging therapies inhibiting platelet receptors, integrins, and proteins involved in platelet activation include the thromboxane inhibitors, ADP receptor antagonists, GP IIb/IIIa inhibitors, and the novel PAR antagonists and adhesion antagonists. *TNF*, tumor necrosis factor; *TXS*, thromboxane A_2 synthase. *Combined thromboxane-receptor antagonists and TXS inhibitors.

From Franchi F, Angiolillo DJ. Novel antiplatelet agents in acute coronary syndrome Nat Rev Cardiol. 2015;12: 30–47; with permission.

Table 2.2 Summary of granule contents for hemostasis and thrombosis.

Granules	Roles	Contents
Alpha (α)	Integral membrane proteins	GP IIb/IIIa, GP1b-IX-V, GP VI, P-selectin
	Coagulants, anticoagulants, and fibrinolytic proteins	Factors V, IX, and XIII; antithrombin; protein S; tissue factor pathway inhibitor; plasminogen; α_2-macroglobulin
	Adhesive proteins	Fibrinogen, vWF, thrombospondin
Dense (δ)	Cations	Ca^{2+}, Mg^{2+}, K^+
	Phosphates	Polyphosphate, pyrophosphate
	Bioactive amines	Serotonin, histamine
	Nucleotides	ADP, ATP, GTP, UTP
Lysosomal (λ)	Enzymes	Acid phosphatase, collagenase, elastase, glucosidase, etc.

ADP, *adenosine diphosphate;* ATP, *adenosine triphosphate;* GP, *glycoprotein;* GTP, *guanosine triphosphate;* UTP, *uridine triphosphate.*

Platelet activation

Platelet plug formation at sites of vascular injury occurs in three stages (see also Chapter 1): (1) the initiation phase involves adhesion of platelets to the subendothelium, (2) the extension phase includes the recruitment of additional platelets and amplification of their response, and (3) the perpetuation phase is characterized by cross-linking and stabilization of the newly formed platelet plug.[1]

Initiation phase

The vascular endothelium normally inhibits platelet activation through multiple mechanisms: ectonucleotidases, which degrade adenosine triphosphate (ATP) and ADP; thrombomodulin, which inactivates thrombin; and the release of prostaglandin (PG) I_2 and nitric oxide (NO).[11] However, upon vessel damage, platelets adhere to the damaged site and aggregate through interactions of platelet receptors with extracellular ligands and soluble proteins. During this initial phase, platelet adhesion at sites of vascular injury is mediated by the binding of vWF to the platelet surface GP Ib-IX-V complex. This interaction allows the binding of exposed collagen to GP VI and GP Ia/IIb receptors on the platelet surface and mediates platelet activation.[12]

Extension phase

After the initial adhesion of platelets to the extracellular matrix, additional platelets are activated and recruited via autocrine and paracrine mediators, including thrombin, ADP, thromboxane A_2 (TXA_2), and epinephrine. During the extension phase, these soluble agonists, which are secreted from activate platelets, play a crucial role for further amplification of platelet activation. G-protein-coupled receptors are the main operating sites of most agonists.[13]

Thrombin is the most potent platelet agonist and activates platelets through PARs and GP Ib-IX-V.[14,15] Although thrombin is generated by the extrinsic pathway of the coagulation cascade, most thrombin is generated on the membrane surface of the activated platelet through the action of prothrombinase, which is composed of the serine protease factor Xa and factor Va.[16] PAR1 and PAR4 are the main receptors for thrombin. However, PAR1 has been suggested to form heterodimers with PAR4 that enhance the cleavage of PAR4.[17] Moreover, PAR1 is more sensitive to low levels of thrombin than PAR4 and cleavage of PAR4 by thrombin takes much longer than cleavage of PAR1.[18] Therefore PAR1 is the most important receptor for thrombin-induced platelet activation, making it a target site for antiplatelet agents.[19]

ADP is a major component of the secreted granules from activated platelets and is an agonist for two G protein-coupled receptors: $G\alpha_q$-coupled $P2Y_1$ contributes to initial activation and is responsible for platelet shape change and $G\alpha_i$-coupled $P2Y_{12}$ decreases cyclic adenosine monophosphate (cAMP) levels, stabilizing the platelet aggregate.[20] Although $P2Y_1$ initiates ADP-induced platelet aggregation, without $P2Y_{12}$ activation, its effects are small and reversible.[21] $P2Y_{12}$ stimulation results in amplification and stabilization of the aggregation response. Thus synergism of the $P2Y_1$ and $P2Y_{12}$ pathways is necessary for ADP-induced platelet activation.[22] In addition, $P2Y_{12}$ and PAR4 have been reported to dimerize, and their interaction promotes PAR signaling, leading to granule release.[23] Because of its key role in platelet activation and amplification processes, the $P2Y_{12}$ receptor has become a key target of antiplatelet therapy. A $P2Y_{12}$ receptor antagonist added to aspirin, known as dual antiplatelet treatment (DAPT), plays a key role in high-risk patients, such as those with acute coronary syndrome (ACS) and those undergoing percutaneous coronary intervention (PCI) with stent implantation.[24-26]

Activated platelets synthesize biologically active lipid species, called eicosanoids such as TXA_2, PGI_2, and PGE_2. The calcium-dependent activation of cytosolic phospholipase A_2 hydrolyzes membrane phospholipids to release arachidonic acid, which are converted into PGs through cyclooxygenase (COX) and lipoxygenase.[27] Although both COX-1 and COX-2 are expressed in young platelets, mature platelets express only COX-1.[28] COX-1 is a housekeeping enzyme that has an important role in producing TXA_2 mostly from platelets.[29] In contrast to constitutively expressed COX-1, COX-2 expression is induced by cytokines, growth factors, and other inflammatory stimuli.[29] Among several PGs synthesized in platelets, TXA_2 is a potent platelet agonist and plays an important role in augmenting platelet activation and promoting thrombus formation by a single TP receptor coupled to both $G\alpha_q$ and $G\alpha_{12/13}$.[30,31] Aspirin irreversibly inhibits platelet aggregation by acetylating the serine residue of COX-1.[28]

Perpetuation phase

Platelet activation stimulates intracellular signaling pathways (increasing Ca^{2+} and decreasing cAMP levels) leading to conformational changes and oligomerization of the integrin $\alpha_{IIb}\beta_3$, which is transformed from its resting low-affinity state to a high-affinity receptor for fibrinogen and vWF.[32,33] During the perpetuation phase, activated $\alpha_{IIb}\beta_3$ allows for the platelet-fibrin-rich plug to stabilize at the site of injury. Because of being the final common pathway, $\alpha_{IIb}\beta_3$ has been extensively evaluated as a target site for antithrombotic therapy and has led to the development of GP IIb/IIIa receptor antagonists.[34]

Antiplatelet pharmacology

This section reviews the mechanism, efficacy, and clinical outcomes of antiplatelet agents most commonly used for treating atheroscleoritc disease manifestations.

Cyclooxygenase 1 inhibitor

Aspirin

The mechanism of aspirin inhibition of PG biosynthesis was discovered in 1971 after it became commercially available because of its anti-inflammatory and analgesic effects in 1899.[35] This particular antiplatelet effect is characteristic of low-dose aspirin, which results from acetylation of serine 529 residue of COX-1 (serine 516 in COX-2 for high-dose aspirin), irreversibly inhibiting access to the COX catalytic site by arachidonic acid.[28,35] Aspirin is absorbed quickly in the upper gastrointestinal tract and achieves peak plasma levels in 30–40 min.[36] Enteric-coated aspirin significantly increases this time to about 3–4 h.[37] The half-life of aspirin in the systemic circulation is only 15–20 min. However, its antiplatelet effects last for the lifespan of the platelet (about 10 days) because platelets cannot generate new COX.[36] Although aspirin irreversibly inhibits platelets, the recovery of platelet function is rapidly occurring within 3–4 days after aspirin discontinuation because COX activity recovers by 10% per day and hemostasis may normalize with more than 20% of COX activity.[38] A maintenance dose (MD) of aspirin 30 mg effectively suppresses platelet TXA_2 after 1 week in healthy individuals.[39] However, a daily aspirin dose of 75–100 mg has also been suggested to be less effective in certain clinical situations (e.g., diabetes mellitus [DM]).[28]

Numerous studies have assessed the role of low-dose aspirin in primary and secondary prevention of CV events. While recent randomized clinical trial and meta-analysis have shown controversial results on the role of aspirin for primary prevention,[40] studies have consistently demonstrated that low-dose aspirin decreases CV events in patients with established CV disease.[41,42] These studies have shown that a low-dose regimen of aspirin provides similar (or even better) efficacy with a lower risk of bleeding compared to high-dose regimens. However, although a once-daily low-dose (75–100 mg) aspirin regimen represents the standard of care and is recommended in practice guidelines,[43–45] the dosage and regimen of aspirin have been the subject of debate and the topic of a number of investigations.

The Clopidogrel and Aspirin Optimal Dose Usage to Reduce Recurrent Events-Seventh Organization to Assess Strategies in Ischemic Syndromes (CURRENT-OASIS 7) study confirmed no significant difference in the rate of CV death, MI, or stroke at 30 days among patients with ACS (n = 25,086) randomly assigned to high-dose (300−325 mg daily) versus low-dose (75−100 mg daily) aspirin (4.2% vs. 4.4%, hazard ratio [HR] 0.97; 95% confidence interval [CI], 0.86−1.09; $P = .61$).[46] The incidence of major bleeding was not significant but minor bleeding was more frequent with high-dose aspirin (5.0% vs. 4.4%, HR 1.13; 95% CI, 1.0−1.27; $P = .04$). Similar findings were observed in a recent analysis from the Longitudinal Assessment of Treatment Patterns and Events after Acute Coronary Syndrome (TRANSLASTE-ACS) registry.[47] Among 10,213 patients with MI who underwent PCI, 63% were discharged on aspirin 325 mg daily and 37% on 81 mg daily. The risk of major adverse CV events (composite of death, MI, stroke, or revascularization) within 6 months was not significantly different between groups (high- vs. low-dose aspirin: 8.2% vs. 9.2%, adjusted HR, 0.99; 95% CI, 0.85−1.17). However, high-dose aspirin was associated with a greater risk of any Bleeding Academic Research Consortium (BARC)-defined bleeding (21.4% vs. 19.5%, adjusted odds ratio [OR], 1.19; 95% CI, 1.05−1.34). More data on the safety and efficacy of low versus high dose aspirin will be obtained from the Aspirin Dosing: A Patient-Centric Trial Assessing Benefits and Long-term Effectiveness (ADAPTABLE) trial comparing a daily dose of aspirin 81 versus 325 mg in 20,000 subjects with established coronary artery disease (CAD).[48]

The regimen of aspirin (once vs. twice daily) has also been questioned. In fact, there are certain clinical scenarios, such as patients with DM, characterized by high platelet turnover rates.[49] Therefore the use of a twice-daily aspirin regimen would allow for more consistent platelet inhibitory effects by inhibiting platelets newly introduced into the blood stream and hence not exposed by aspirin if given once daily due to its very short plasma half-life. While pharmacodynamic studies have shown a twice-daily regimen to achieve optimized platelet inhibitory effects compared to a once-daily regimen in patients with DM, the clinical implications of these findings warrant investigation.[50] The ongoing Aspirin Twice a Day in Patients with Diabetes and Acute Coronary Syndrome (ANDAMAND) (NCT02520921) trial will clarify the net benefit of twice-daily administration of low-dose aspirin.

P2Y$_{12}$ inhibitors

ADP is the agonist of the P2Y$_{12}$ receptor. The P2Y$_{12}$ receptor is coupled to $G\alpha_{i2}$ causing inhibition of cAMP, which plays a critical role in keeping platelets in a resting state.[20] The cAMP-dependent protein kinase A mediates phosphorylation of vasodilator-stimulated phosphoprotein (VASP).[51] Decreasing cAMP levels promotes dephosphorylation of VASP, which leads to changes in the GP IIb/IIIa receptor to its activated state. $G\alpha_i$ stimulates platelet activation not only by decreasing cAMP levels but also via cAMP-independent pathway.[52] In particular, $G\alpha_i$ interacts with the activation of phosphatidylinositol-3-kinase (PI3K) as well as RAS-related protein b (Rap1b), which leads to GP IIb/IIIa receptor activation, granule release, and stabilization of the platelet aggregate. In addition, P2Y$_{12}$ makes a dimer with

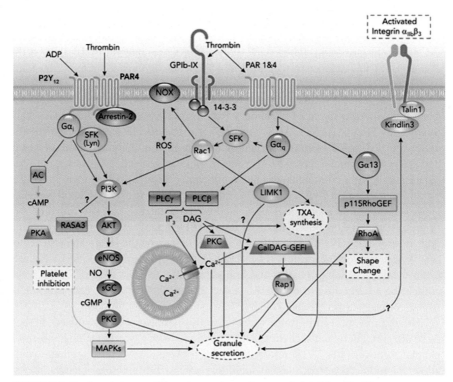

FIGURE 2.2 Synergy between different receptor pathways of platelet activation.

Platelet agonists such as adenosine diphosphate (ADP) and thrombin have more than one platelet receptor for optimal platelet activation, which requires synergy or cooperativity between different receptor pathways. Cooperativity of $G\alpha_q$-coupled P2Y$_1$ and $G\alpha_i$-coupled P2Y$_{12}$ pathway is necessary for ADP-induced platelet activation. Thrombin has at least three receptors including protease-activated receptor (PAR) 1, PAR4, and glycoprotein (GP) Ib-IX in human platelets. Moreover, interactions between PARs and P2Y$_{12}$ receptors have been reported, which may produce a synergistic effect of platelet activation through various pathways. *AC*, adenylyl cyclase; *DAG*, diacylglycerol; *IP$_3$*, inositol 1,4,5-tirphosphate; *NOX*, NADPH oxidase; *p115 RhoGEF*, p115 Rho guanine nucleotide exchange factor; *PI3K*, phosphatidylinositol-3-kinase; *PKA*, cAMP-dependent protein kinase; *Rap 1b*, RAS-related protein 1b; *ROS*, reactive oxygen species; *TXA$_2$*, thromboxane A$_2$.

From Estevez B, Du X. New concepts and mechanism of platelet activation signaling. Physiology 2017;32: 162–177; with permission.

PAR4 and their interaction leads to platelet granule release (Fig. 2.2).[9,23] Because of its key role in platelet activation and amplification processed, the P2Y$_{12}$ receptor has been regarded as an ideal target for an antiplatelet agent. P2Y$_{12}$ receptor antagonists are generally classified as thienopyridines (ticlopidine, clopidogrel, and prasugrel) and nonthienopyridines (ticagrelor and cangrelor). The first-generation P2Y$_{12}$

inhibitor ticlopidine is no longer utilized because of its side effects, in particular, bone marrow suppression.[53] In this section, we review the currently available P2Y$_{12}$ inhibitors: clopidogrel, prasugrel, ticagrelor, and cangrelor (Table 2.3).

Clopidogrel

Clopidogrel is a prodrug that requires metabolic transformation to exert its antiplatelet effects.[54] After intestinal absorption through P-gp encoded by ABCB1, clopidogrel undergoes hepatic activation by the cytochrome P450 (CYP) system (Fig. 2.3). Approximately 85% of absorbed clopidogrel is hydrolyzed by esterases into an inactive metabolite and only 15% is available for hepatic metabolism, which requires two sequential oxidative steps.[54] In the first step, clopidogrel is catalyzed by the CYP system to 2-oxo-clopidogrel, followed by ring opening to the corresponding highly unstable sulfenic acid or thiol, which covalently binds to the P2Y$_{12}$ receptor. CYP1A2, CYP2B6, CYP2C9, CYP2C19, and CYP3A4 all contribute to clopidogrel

Table 2.3 Pharmacologic characteristics of P2Y$_{12}$ antagonists.

	Clopidogrel	Prasugrel	Ticagrelor	Cangrelor
Classification	Thienopyridine	Thienopyridine	CPTP	ATP analogue
Bioavailability	Prodrug	Prodrug	Direct-acting	Direct-acting
Receptor blockade	Irreversible	Irreversible	Reversible	Reversible
Onset of action	2–8 h	30 min to 4 h	30 min to 4 h	3–6 min
Offset of action	5–7 days	7–10 days	3–5 days	30–60 min
Route of administration	Oral	Oral	Oral	Intravenous
Dose for PCI (loading/maintenance)	600 mg/75 mg qd	60 mg/10 mg qd	180 mg/90 mg bid	30 µg/kg/4 µg/kg/min
Drug interactions	CYP2C19 inhibitors	N/A	CYP3A4/5 inhibitor or inducer	Clopidogrel, prasugrel
Indications	ACS and stable CAD undergoing PCI	ACS undergoing PCI, (caution for patients aged ≥75 years or weighing <60 kg)[a]	ACS (irrespective of management)	Periprocedural thrombotic events without previous P2Y$_{12}$ antagonists and GPIs
Contraindication	Active bleeding	Active bleeding, history of TIA or stroke	Active bleeding, ICH, severe hepatic impairment	Active bleeding

ACS, *acute coronary syndrome;* ATP, *adenosine triphosphate;* CAD, *coronary artery disease;* CPTP, *cyclopentyltriazolopyrimidine;* CYP, *cytochrome P450;* GPI, *glycoprotein IIb/IIIa inhibitor;* ICH, *intracranial hemorrhage;* N/A, *not applicable;* PCI, *percutaneous coronary intervention;* qd, *every day;* TIA, *transient ischemic attack.*
[a] *Consider 5 mg qd.*

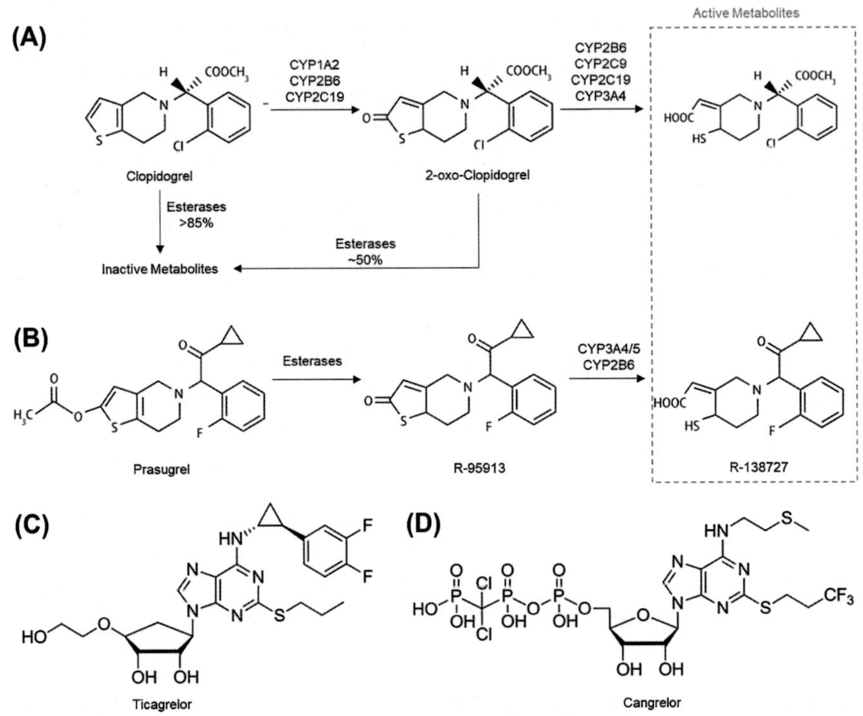

FIGURE 2.3 Chemical structures and active metabolites of P2Y$_{12}$ receptor inhibitors.

(A) Clopidogrel required a two-step process mediated by cytochrome P450 (CYP) enzymes to exert an antiplatelet effect. Oxidation is catalyzed by CYP1A2, CYP2B6, and CYP2C19, respectively, to 2-oxo-clopidogrel and its sulfoxide derivative followed by ring opening to the corresponding highly unstable sulfenic acid, which covalently binds to the receptor protein. (B) Prasugrel is activated by esterase in the intestine and blood serum, which is then converted to its metabolite R-138727 by CYP enzymes. This one-step oxidation of prasugrel attributes less resistance in patients and fewer drug interactions compared with clopidogrel. (C) The nucleotide-derived ticagrelor is an orally active, reversible, and competitive P2Y$_{12}$ receptor antagonist. (D) Cangrelor is an intravenous analogue of adenosine triphosphate with a rapid offset of action. Both ticagrelor and cangrelor are active metabolites that lack many of the disadvantages associated with prodrugs such as clopidogrel and prasugrel.

metabolism.[54] However, CYP2C19 is significantly involved in both oxidative steps.[54] This is the reason for the presence of impaired clopidogrel-induced antiplatelet effects in the presence of genetic polymorphisms associated with reduced enzyme activity or drugs interfering with CYP2C19.[55]

A loading dose (LD) of 600 mg is typically used in PCI-treated patients, as this is more effective in suppressing platelet aggregation than a 300 mg LD. However, an LD higher than 600 mg is not associated with enhanced platelet inhibition due to limited intestinal absorption.[56] The daily MD is usually 75 mg, which requires 3−7 days to reach steady-state platelet inhibition. An MD of 150 mg daily is associated with reduced platelet reactivity and enhanced platelet inhibition compared to 75 mg daily.[57] However, this MD regimen is not recommended or utilized in clinical practice.

Compared with aspirin, clopidogrel monotherapy decreased the risk of ischemic events (composite of vascular death, MI, or stroke) in patients with atherosclerotic vascular disease (5.32% vs. 5.83%, relative risk [RR] 8.7%; 95% CI, 0.3−16.5; $P = .043$) without increasing bleeding risk and with fewer gastrointestinal side effects.[58] When added to aspirin, clopidogrel has resulted in better clinical outcomes across the spectrum of patients with ACS and in the setting of PCI. In the Clopidogrel in Unstable Angina to Prevent Recurrent Ischemic Events (CURE) trial, 12,562 patients with ACS including unstable angina and non-ST-elevation myocardial infarction (NSTEMI) were randomized to aspirin and clopidogrel (300 mg LD followed by 75 mg MD) or aspirin alone for 3−12 months. When compared with aspirin alone, clopidogrel with aspirin reduced the rate of the composite endpoint of MI, stroke, or CV death by 20% (9.3% vs. 11.4%, RR 0.8; 95% CI, 1.13−1.67; $P < .001$).[24] Patients on clopidogrel plus aspirin had significantly more major bleeding than on aspirin alone (3.7% vs. 2.7%, RR 1.38; 95% CI, 1.13−1.67; $P = .001$), but fatal bleeding did not differ between the two groups (2.2% vs. 1.8%, RR 1.21; 95% CI, 0.95−1.56; $P = .13$).[24]

The adjunctive benefit of clopidogrel to aspirin was also evaluated in two trials of patients with ST-elevation myocardial infarction (STEMI). The Clopidogrel and Metoprolol in Myocardial Infarction Trial (COMMIT) randomized 45,852 patients with suspected STEMI within 24 h of onset to clopidogrel and aspirin or aspirin alone.[59] Clopidogrel showed a highly significant reduction in all-cause mortality (clopidogrel vs. placebo: 7.5% vs. 8.1%, OR, 0.93; 95% CI, 0.87−0.99; $P = .03$) without excessive risk of bleeding (0.58% vs. 0.55%, $P = .59$).[59] The Clopidogrel as Adjunctive Reperfusion Therapy-Thrombolysis in Myocardial Infarction (CLARITY-TIMI 28) trial randomized 3491 patients treated with thrombolysis within 12 h of STEMI to clopidogrel 300 mg followed by 75 mg daily or placebo. Clopidogrel significantly improved the patency rate of the infarct-related artery and reduced death or recurrent MI (15.0% vs. 21.7%, $P < .001$) without increasing bleeding risk.[60]

The long-term effect of clopidogrel in patients undergoing PCI was studied in the Clopidogrel for the Reduction of Events during Observation (CREDO) trial.[61] In the CREDO trial, 2116 patients were randomized into two groups: pretreatment (before PCI) with a 300 mg clopidogrel LD and 75 mg daily MD for 12 months versus clopidogrel 75 mg daily without loading for 1 month and then placebo for 11 months. Although clopidogrel did not show any clinical benefit at 28 days (loading vs. no loading: 6.8% vs. 8.3, $P = .23$), clopidogrel therapy for 12 months showed significant relative risk reduction (RRR) in the combined risk of death, MI, or stroke (RRR, 26.9%; 95% CI, 3.9%−44.4%; $P = .02$). The risk of major bleeding was not significantly increased with clopidogrel (8.8% vs. 6.7%, $P = .07$). The

Clopidogrel for High Atherothrombotic Risk and Ischemic Stabilization, Management, and Avoidance (CHARISMA) trial studied the efficacy of clopidogrel plus aspirin compared with aspirin alone in patients with clinically evident atherothrombotic disease manifestation or multiple CV risk factors.[62] Although clopidogrel plus aspirin did not significantly reduce the rate of the primary efficacy endpoint including MI, stroke, or death from CV causes (6.8% vs. 7.3%, RR 0.93; 95% CI, 0.83–1.05; $P = .22$) in the overall trial, the rate of the primary endpoint in patients with symptomatic atherothrombosis was significantly lower with clopidogrel plus aspirin compared with aspirin alone (6.9% vs. 7.9%, RR 0.88; 95% CI, 0.77–0.998; $P = .046$). On the contrary, among patients with multiple risk factors, clopidogrel plus aspirin had a higher rate of CV death than aspirin alone (3.9% vs. 2.2%, $P = .01$). The rate of severe bleeding was not significantly different (1.7% vs. 1.3%, RR 1.25: 95% CI, 0.97–1.61; $P = .09$) but clopidogrel had a higher rate of moderate bleeding (2.1% vs. 1.3%, RR 1.62: 95% CI, 1.27–2.08; $P < .001$).[62]

Despite the clinical benefits of clopidogrel, it is limited by variability in its antiplatelet effects and a considerable number of patients persist with high platelet reactivity (HPR).[63] Importantly, HPR has shown to be a predictor of thrombotic complications.[63] The antiplatelet effect of clopidogrel is limited by the generation of an active metabolite through intestinal absorption and hepatic CYP450 system.[54] Thus a number of factors in drug absorption and metabolism can contribute to variability in clopidogrel effects. These have included but not limited to clinical (age, diabetes, body mass index, drug interactions), genetic (CYP2C19 genotypes), and cellular (upregulation of platelet signaling pathways) factors[55,63] Higher doses of clopidogrel decrease but cannot eliminate this variability.[64] Important contributors to impaired clopidogrel response have led to boxed warning including CYP2C19 loss-of-function polymorphisms and concomitant use of proton pump inhibitors (PPIs).[55] CYP2C19 loss-of-function polymorphisms have been associated with reduced clopidogrel-induced antiplatelet effects and increased risk of thrombotic complications, particularly in patients undergoing PCI.[65] Although PPIs have shown to affect clopidogrel effects, the clinical impact of these findings has been less consistent.[66] The limitations of clopidogrel have prompted the development of newer P2Y$_{12}$ receptor antagonists.

Prasugrel

Prasugrel is a prodrug that requires a single-step oxidation process involving hepatic CYP to generate its active metabolite. Hydrolysis of the acetate ester portion of prasugrel by human carboxylesterase 2 during absorption leads to a thiolactone metabolite and then a single-step CYP oxidation allows for its conversion to the active metabolite (Fig. 2.3).[67] CYP3A4/5 and CYP2B6 play a major role in this oxidation process.[68] Although the active metabolites of prasugrel and clopidogrel have the same potency, prasugrel has a faster onset of action, more potent inhibition, and lower interindividual response variability than clopidogrel because of the more efficient metabolic conversion of prasugrel.[69,70]

The antiplatelet potency of prasugrel is at least 10 times higher than that of clopidogrel.[70] Both the 60 mg LD and 10 mg MD of prasugrel have shown more prompt and potent platelet inhibition compared with the 300 mg LD and 75 mg MD of clopidogrel.[71] In the Trial to Assess Improvement in Therapeutic Outcomes by Optimizing Platelet Inhibition with Prasugrel-Thrombolysis in Myocardial Infarction (TRITON-TIMI) 38, prasugrel (60 mg LD and 10 mg MD) was associated with fewer CV events (death from CV causes, nonfatal MI, or stroke) than clopidogrel (300 mg LD and 75 mg MD) in patients with ACS undergoing PCI (9.9% vs. 12.1%, HR 0.81; 95% CI, 0.73–0.90, $P < .001$), driven by a reduction in MI.[25] However, in the Targeted Platelet Inhibition to Clarify the Optimal Strategy to Medically Manage Acute Coronary Syndromes (TRILOGY ACS) trial, which enrolled medically managed patients with ACS who did not undergo revascularization, the primary composite endpoint including death from CV causes, MI, or stroke was not significantly different between prasugrel and clopidogrel (13.9% vs. 16.0%, HR, 0.91; 95% CI, 0.79–1.05, $P = .21$).[72] In TRITON-TIMI 38, prasugrel was associated with harm in patients with a past history of stroke and had neutral effect in elderly patients (age ≥ 75 years) and in those with low body weight (< 60 kg). For patients aged ≥ 75 years or weighing < 60 kg, a 5 mg MD has been suggested but the clinical efficacy and safety of this dose is limited.[73] To evaluate the effects of pretreatment with prasugrel, the Comparison of Prasugrel at the Time of Percutaneous Coronary Intervention or as Pretreatment at the Time of Diagnosis in Patients with Non-ST Elevation Myocardial Infarction (ACCOAST) trial was designed.[74] In the ACCOAST trial, 4033 patients with NSTEMI were randomly assigned to pretreatment (a 30 mg LD before PCI) or placebo. In patients undergoing PCI, the pretreatment group received an additional 30 mg of prasugrel while the placebo group received 60 mg of prasugrel. The rate of the primary endpoint (composite of CV death, MI, stroke, urgent revascularization, or GP IIb/IIIa inhibitor rescue therapy through day 7) was not significantly different between the two groups (HR with pretreatment, 1.02; 95% CI, 0.84–1.25; $P = .81$). However, TIMI major bleeding was higher in the pretreatment group (2.6% vs. 1.4%, HR 1.09; 95% CI, 1.19–3.02; $P = .006$).

Ticagrelor

The nucleoside analogue ticagrelor is an orally active, competitive, and reversible antagonist of the $P2Y_{12}$ receptor (Fig. 2.3).[75] Interestingly, ticagrelor may act in some conditions as an inverse agonist that stabilizes the $P2Y_{12}$ receptor into an inactivate state and also promote inhibition of ADP-induced Ca^{2+} release.[76,77] Ticagrelor also exerts effects beyond $P2Y_{12}$ antagonism by inhibiting the equilibrative nucleoside transporter 1 (ENT1) on red blood cells, leading to the accumulation of extracellular adenosine and activation of the Gs-coupled adenosine A_{2A} receptor.[77] This contributes to an increase in cAMP levels and phosphorylation of VASP. These off-target effects can contribute to ticagrelor's efficacy. Because of its metabolism through hepatic CYP3A4/5, coadministration of ticagrelor with CYP3A4/5 inducers decreases plasma concentration and potentially increases the risk of thrombotic events; conversely, CYP3A4/5 inhibitors increase ticagrelor's concentration and potentially augment the risk of bleeding.[10]

Ticagrelor is rapidly absorbed with a peak plasma concentration at 90 min and, because of its half-life of 8—12 h, requires a twice-daily dosing regimen.[78] Ticagrelor has more rapid, potent, and predictable antiplatelet effect with a relatively faster offset of action than clopidogrel.[79] Ticagrelor 180 mg LD and 90 mg twice daily MD provide higher platelet inhibition than clopidogrel 600 and 75 mg daily MD. Because of its rapid offset with ticagrelor, inhibition of platelet aggregation with ticagrelor on day 3 after the last dose is comparable with that for clopidogrel at day 5.

Unlike the TRITON-TIMI 38 trial in which prasugrel was selectively tested in patients with ACS undergoing PCI, the Platelet Inhibition and Patient Outcomes (PLATO) study was conducted using ticagrelor in patients with ACS irrespective of their management (invasive and noninvasive).[26] In PLATO, ticagrelor with aspirin showed significant clinical benefit in patients with ACS with or without ST elevation, resulting in lower rates of the primary endpoint (a composite of death from vascular causes, MI, and stroke) at 12 months compared with clopidogrel and aspirin (9.8% vs. 11.7%; HR 0.84, 95% CI, 0.77—0.92, $P < .001$).[26] These findings were driven by a reduction in CV mortality and MI. The rates of major bleeding as defined in the trial did not differ between ticagrelor and clopidogrel (11.6% vs. 11.2%, HR 1.04; 95% CI, 0.95—1.13; $P = .43$). However, this also resulted in higher rates of non—coronary artery bypass graft (CABG)-related bleeding (4.5% vs. 3.8%; HR 1.19, 95% CI, 1.02—1.38, $P = .03$).[26] The Prevention of Cardiovascular Events in Patients with Prior Hear Attack Using Ticagrelor Compared to Placebo on a Background of Aspirin-Thrombolysis in Myocardial Infarction 54 (PEGASUS-TIMI 54) trial investigated the effect of ticagrelor beyond 1 year after an ACS. Thus 21,162 patients who had had an MI 1—3 years earlier were randomized to ticagrelor (two dosing regimens: 60 and 90 mg bid) and aspirin or aspirin alone for a median of 33 months.[80] Both ticagrelor doses decreased, as compared with placebo, the risk of the primary composite endpoint of CV death, MI, or stroke (60 mg vs. placebo, 7.77% vs. 9.04%, HR 0.84; 95% CI, 0.74—0.95; $P = .004$; 90 mg vs. placebo, 7.85% vs. 9.04%, HR 0.85; 95% CI, 0.75—0.96; $P = .008$). Although TIMI major bleeding occurred more frequently in both dose groups than placebo (60 mg vs. placebo, 2.3% vs. 1.06%, HR 2.32; 95% CI, 1.68—3.21; $P < .001$; 90 mg vs. placebo, 2.6% vs. 1.06%, HR 2.69; 95% CI, 1.96—3.7; $P < .001$), the rate of fatal bleeding or nonfatal intracranial hemorrhage did not differ between ticagrelor dose group and placebo (60 mg vs. placebo, 0.71% vs. 0.6%, HR 1.2; 95% CI, 0.73—1.97; $P = .47$; 90 mg vs. placebo, 0.63% vs. 0.6%, HR 1.22; 95% CI, 0.74—2.01; $P = .43$).[80]

The Intracoronary Stenting and Antithrombotic Regimen: Rapid Early Action for Coronary Treatment 5 (ISAR-REACT 5) trial randomized 4018 patients with ACS undergoing PCI to ticagrelor versus prasugrel. Compared with prasugrel, the composite events of death, MI, or stroke were significantly higher in the ticagrelor group (6.9% vs. 9.3%, HR 0.64; 95% CI, 0.3—0.91, $P = .006$).[81]

The Effect of Ticagrelor on Health Outcomes in Diabetes Mellitus patients Intervention Study (THEMIS) randomized 19,220 patients who had stable CAD and type II DM to ticagrelor plus aspirin or placebo plus aspirin.[82] Ticagrelor plus aspirin reduced the ischemic CV events (CV death, MI, or stroke) compared with placebo plus aspirin (7.7% vs. 8.5%, HR 0.90; 95% CI, 0.81–0.99, $p = .04$), whereas the incidence of TIMI major bleeding was significantly higher (2.2% vs. 1.0%, HR 2.32; 95% CI, 1.82–2.94, $P < .001$).[82]

In light of the increased bleeding, in particular gastrointestinal, associated with the use of aspirin and the results of pharmacodynamics investigations suggesting that aspirin provides limited additional antiplatelet effects, a number of studies have evaluated the impact of dropping aspirin on a background of potent $P2Y_{12}$ inhibition, in particular ticagrelor, in patients undergoing PCI. GLOBAL LEADERS was the first large-scale trial to assess the clinical impact of dropping aspirin. The trial randomized 15,968 patients undergoing PCI for stable CAD or ACS to an aspirin-free approach (75–100 mg aspirin daily plus 90 mg ticagrelor twice daily for 1 month, followed by 23 months of ticagrelor alone) versus standard of care (75–100 mg aspirin daily plus 75 mg clopidogrel daily in patients with stable CAD or 90 mg ticagrelor twice daily in patients with ACS for 12 months, followed by aspirin monotherapy for 12 months).[83] At 2-year follow-up, the primary composite event of all-cause death or new Q-wave MI was not different between experimental and control groups (3.81% vs. 4.37%, RR 0.87; 95% CI, 0.75–1.01; $P = .073$), irrespective of clinical presentation. BARC grade 3 or 5 events did not differ significantly between experimental and control groups (2.04% vs. 2.12%, RR 0.97; 95% CI, 0.78–1.20; $P = .77$).[83] Different findings were observed in other trials. In the SMART-CHOICE trial, 2993 patients undergoing PCI were randomized to $P2Y_{12}$ inhibitor monotherapy (aspirin plus a $P2Y_{12}$ inhibitor for 3 months, followed by a $P2Y_{12}$ inhibitor alone) versus DAPT (aspirin plus a $P2Y_{12}$ inhibitor for at least 12 months).[84] The primary composite endpoint of all-cause death, MI, or stroke at 12 months was similar between the 2 groups (2.9% vs. 2.5%, difference, 0.4% [one-sided 95% CI, 1.3%]; $P = .007$ for noninferiority), with a greater frequency of BARC 2 to 5 bleeding observed in the DAPT group (2.0% vs. 3.4%, HR, 0.36; 95% CI, 0.36–0.92; $P = .02$).[84] The Short and Optimal Duration of Dual Antiplatelet Therapy (STOPDAPT) 2 trial randomized 3045 patients undergoing PCI to 1-month DAPT, followed by clopidogrel monotherapy or 12-month DAPT (aspirin plus clopidogrel).[85] The major CV endpoint including CV death, MI, stent thrombosis, or stroke was similar between the 2 groups (1.96% vs. 2.51%, absolute difference, −0.55%; 95% CI, −1.62 to 0.52; $P = .005$ for noninferiority), while the bleeding endpoint (composite of TIMI major or minor bleeding) was significantly higher with 12-month DAPT (0.41% vs. 1.54%; HR, 0.26; 95% CI, 0.11–0.64; $P = .004$). In the Ticagrelor with Aspirin or Alone in High-Risk Patients After Coronary Intervention (TWILIGHT) trial, among patients who underwent PCI, after a 3-month period of DAPT with aspirin and ticagrelor, ticagrelor monotherapy decreased the incidence of clinically relevant bleeding compared with ticagrelor plus aspirin at 12 months (4.0% vs. 7.1%, HR 0.56; 95% CI, 0.45–0.68, $P < .001$) without increasing the risk of death, MI, or stroke (3.9% vs. 3.9%, HR 0.99; 95% CI, 0.78–1.25).[86]

In addition to bleeding, ticagrelor is also associated with a series of nonhemorrhagic adverse effects, including dyspnea, ventricular pauses, and increased serum levels of creatinine and uric acid.[87] In particular, dyspnea has been reported to occur in 15%−29% of patients and represents a contributing cause to ticagrelor cessation.[26,81,82,87] However, pulmonary function parameters in patients with ACS were not different between ticagrelor and clopidogrel after a mean treatment duration of 30 days.[88]

Cangrelor

Cangrelor is an intravenous analogue of ATP and a reversible antagonist of the P2Y$_{12}$ receptor (Fig. 2.3).[89] Cangrelor is metabolically unstable and thus has a short half-life (3−6 min) resulting in a rapid offset of action (<60 min).[90,91] After intravenous administration as a 30 µg/kg bolus followed by a 4 µg/kg/min infusion, the peak concentration of cangrelor was achieved within 2 min and near-full recovery of platelet function was shown within 60 min after termination of infusion.[90] Given that it is not renally excreted, no dose modification is required based on renal function. When transitioning from cangrelor to an oral P2Y$_{12}$ inhibitor, an LD of clopidogrel and prasugrel should be administered at the end of cangrelor infusion to allow for the wash out of cangrelor and prevent it from blocking the binding site of the active metabolite of thienopyridines on the P2Y$_{12}$ receptor.[92,93] In contrast, ticagrelor can be used before, during, or after the cangrelor infusion without drug-drug interaction.[93,94]

In the Cangrelor versus Standard Therapy to Achieve Optimal Management of Platelet Inhibition (CHAMPION)-PHOENIX trial, patients were randomized to cangrelor or a 300−600 mg LD of clopidogrel. The composite of death, MI, ischemia-driven revascularization, or stent thrombosis was reduced by cangrelor at 48 h (4.7% vs. 5.9%, OR 0.78; 95% CI, 0.66−0.93; $P = .005$) without increasing the risk of major bleeding.[95] Cangrelor can be a useful antiplatelet agent especially in patients unable to use oral P2Y$_{12}$ antagonists (shock, intubation, vomiting, etc.), in patients requiring fast and reliable platelet inhibition (e.g., STEMI), and in patients not pretreated with a P2Y$_{12}$ inhibitor.[96] In the Cangrelor and Crushed Ticagrelor in STEMI Patients Undergoing Primary Percutaneous Coronary Intervention (CANTIC) trial, cangrelor showed faster platelet inhibition after bolus at 5 min than crushed ticagrelor 180 mg LD, an effect that was sustained during infusion.[97]

The rapid reversal of cangrelor has also made this a suitable drug for bridging patients requiring surgery but who still require P2Y$_{12}$ inhibition. The Bridging Antiplatelet Therapy with Cangrelor in Patients Undergoing Cardiac Surgery (BRIDGE) trial identified a dose of cangrelor of 0.75 µg/kg/min that was then applied as a bridging strategy in patients undergoing CABG after discontinuation of an oral P2Y$_{12}$ inhibitor.[98] Overall, the study showed sustained antiplatelet effects during cangrelor infusion and rapid return to baseline platelet function within 6 h of treatment discontinuation prior to surgery.

Thrombin receptor antagonists

Thrombin, a serine protease, facilitates thrombosis, hemostasis, and inflammation via multiple actions including platelet activation, protein C activation, and conversion of fibrinogen to fibrin.[16] Thrombin is the most potent agonist for platelet activation via PAR1, PAR4, and GP Ib-IX.[14,15] Thrombin binds and cleaves PAR1 and PAR4 to expose a new NH_2-terminal sequence, which serves as an internal ligand to interact and activate receptor signaling.[99] PAR1 is also expressed by endothelial and vascular smooth cells and is associated with inflammation, vascular transcriptional activation, vascular smooth muscle cell migration, and proliferation.[16] Thrombin plays a direct role in the final step of the coagulation cascade by converting fibrinogen to fibrin.[14,16] Platelet activation by PAR1 mainly contributes to thrombosis but not hemostasis.[14] In addition, because of maintaining the fibrin-generating and protein C functions of thrombin, a direct antagonist of PAR1 is an attractive candidate for reducing ischemic events without affecting hemostasis, hence bleeding.

Vorapaxar

Vorapaxar is a tricyclic 3-phenyl pyridine analogue that is synthesized based on the natural product of himbacine.[100] It is an orally active, highly selective, and competitive PAR1 antagonist. After administration, vorapaxar is rapidly absorbed with >90% bioavailability, slowly metabolized by CYP3A4, and eliminated mainly through the feces, while less than 5% is excreted by urine.[101,102] A single 40 mg dose completely inhibits thrombin receptor activating peptide (TRAP)-induced platelet aggregation within 1 h and continues over 3 days.[102] The recovery of platelet function following the 20 and 40 mg LD occurs gradually over a 4- to 8-week period (half-life ~311 h). Complete inhibition of TRAP-induced platelet aggregation is sustained with a 2.5 mg once-daily MD.[103] Despite being a reversible antagonist, the long half-life of vorapaxar makes its effects functionally irreversible, allowing for consistent platelet inhibition. Because vorapaxar is a selective platelet PAR1 inhibitor, coagulation and clotting time are not affected.[104]

The Thrombin-Receptor Antagonist Vorapaxar in Acute Coronary Syndromes (TRACER) trial assessed the efficacy and safety of vorapaxar added on top of aspirin and/or clopidogrel therapy in patients with non-ST-segment elevation ACS.[19] Patients were randomized to vorapaxar (40 mg LD and 2.5 mg daily MD) or placebo. Although vorapaxar did not significantly reduce the primary composite endpoint of CV death, MI, stroke, recurrent ischemia with hospitalization, or urgent coronary revascularization (18.5% vs. 19.9%, HR, 0.92; 95% CI, 0.85−1.01; $P = .07$), it significantly reduced the secondary composite endpoint including CV death, MI, or stroke (14.7% vs. 16.4%, HR, 0.89; 95% CI, 0.81−0.98; $P = .02$). These data were overall driven by a reduction in MI (11.1% vs. 12.5%, HR, 0.88; 95% CI, 0.79−0.98; $P = .02$). However, vorapaxar significantly increased the risk of major bleeding (7.2% vs. 5.2%, HR, 1.35; 95% CI, 1.16−1.58; $P < .001$). The Thrombin-Receptor Antagonist in Secondary Prevention of Atherothrombotic Ischemic Events (TRA 2P)-TIMI 50 trial evaluated the efficacy and safety of vorapaxar in addition to standard antiplatelet therapy for secondary prevention in subjects with a history of MI, ischemic stroke, and peripheral arterial disease (PAD).[105] Patients were randomized to receive either vorapaxar 2.5 mg daily (without an LD) or placebo. At 3 years, vorapaxar decreased the primary endpoint (a composite of CV death, MI, stroke) compared to placebo (9.3% vs. 10.5%, HR, 0.87; 95% CI, 0.8−0.94; $p < .001$). The benefit of vorapaxar in the prevention of ischemic events was most apparent in patients with a history of MI (8.1% vs. 9.7%, HR, 0.8; 95% CI, 0.72−0.89; $P < .001$).[106] In addition, vorapaxar significantly reduced the incidence of hospitalization for acute limb ischemia (2.3% vs. 3.9%, $P = .006$) and revascularization of PAD (18.4% vs. 22.2%, $P = .017$).[107] Similar to the TRACER trial, vorapaxar increased the risk of the Global Use of Strategies to Open Occluded Coronary Arteries (GUSTO) moderate or severe bleeding (4.2% vs. 2.5%, HR, 1.66; 95% CI, 1.43−1.93; $P = .001$).[105] In the United States and Europe, vorapaxar was approved for the reduction of thrombotic CV events in patients with a prior MI or PAD and should be used in addition to standard-of-care antiplatelet therapy.[108] Because of the observed increase in intracranial hemorrhage in patients with a prior cerebrovascular event in the TRACER and TRA-2P trials, vorapaxar is contraindicated in these subjects.

Glycoprotein IIb/IIIa inhibitors

The most abundant and important platelet integrin is GP IIb/IIIa, which mediates the final step of platelet aggregation by serving as a receptor for fibrinogen and VWF and is required for stable platelet adhesion to the vascular wall and platelet aggregation.[9] Following platelet activation, intracellular signals transform the resting GP IIb/IIIa into a high-affinity conformational state and this process is known as inside-out signaling. In addition, GP IIb/IIIa mediates an outside-in signaling pathway thus further enhancing platelet activation and thrombus formation.[9] Because GP IIb/IIIa is the final pathway of platelet aggregation, inhibitors of GP IIb/IIIa have potent antiplatelet effects. Currently used antagonists are administered parentally and classified into two groups: antibodies (abciximab) and synthetic or small molecules (eptifibatide, tirofiban) (Table 2.4).[96] Although GP IIb/IIIa antagonists have shown to significantly reduce periprocedural thrombotic complications particularly in high-risk PCI, their potent antiplatelet effect is associated with increased bleeding risk. Thus contemporary practical use of GP IIb/IIIa is limited to bailout settings.[34,96]

Abciximab

After a murine monoclonal antibody (known as 7E3) against GP IIb/IIIa was originally produced, the Fc fragment was cleaved to reduce its immunogenicity and then a chimeric antibody fragment (c 7E3 Fab) called abciximab was developed.[109,110] Abciximab is a large molecule that exhibits a strong affinity for GP IIb/IIIa and irreversibly binds to the receptor. The recommended dose is 0.25 mg/kg bolus intravenous followed by 0.125 μg/kg/min for 12 h, with no need for renal adjustment.[111]

Table 2.4 Pharmacologic characteristics of GP IIb/IIIa inhibitors.

	Abciximab	Eptifibatide	Tirofiban
Molecule	Fab 7E3, chimeric	Cyclic heptapeptide	Nonpeptide tyrosine derivative
Affinity to GP IIb/IIIa	+++	+	++
Binding	Noncompetitive	Competitive	Competitive
Half-life	10–24 h	2.5 h	2 h
Dose for PCI			
Bolus	0.25 mg/kg (10–60 min)	180 µg/kg (10 min) + 180 µg/kg	25 µg/kg (over 3 min)
Infusion	0.125 µg/kg/min (for 12 h)	2 µg/kg/min (up to 18 h)	0.15 µg/kg/min (up to 18 h)
Renal adjustment[a]			
Bolus	N/A	180 µg/kg	12.5 µg/kg (over 3 min)
Infusion	N/A	1 µg/kg/min (up to 18 h)	1.75 µg/kg/min (up to 18 h)

Fab, antigen-binding fragment; GP, glycoprotein; N/A, not applicable; PCI, percutaneous coronary intervention.
[a] Consider dose adjustment of eptifibatide when creatinine clearance (CrCl) is less than 50 mL/min or of tirofiban when CrCl is less than 60 mL/min.

The bolus dose of 0.25 mg/kg showed >80% receptor blockade and decreased platelet aggregation to less than 20% compared with baseline.[110] Its plasma half-life is initially less than 10 min and a second-phase half-life is about 30 min. However, its elimination from plasma is slow with a half-life of 10–24 h because of avid platelet binding. Therefore it has a biologic half-life of up to 7 days, and platelet-associated abciximab can be detected for more than 14 days after discontinuation of infusion.[111]

The Intracoronary Stenting and Antithrombotic Regimen: Rapid Early Action for Coronary Treatment (ISAR-REACT) trial showed no benefit of abciximab on the primary endpoint of death, MI, or urgent target vessel revascularization (TVR) in patients undergoing elective PCI after pretreatment with clopidogrel 600 mg at 30 days (abciximab vs. placebo, 4% vs. 4%, RR 1.05; 95% CI, 0.69–1.59; $P = .82$).[112] Although there were no differences in major bleeding, thrombocytopenia occurred more frequently in the abciximab group. The ISAR-REACT 2 trial enrolled 2022 patients with ACS pretreated at least 2 h before PCI with clopidogrel 600 mg, in which abciximab decreased the composite of death, MI, and urgent TVR at 30 days compared with placebo (8.9% vs. 11.9%, RR 0.75; 95% CI, 0.58–0.97; $P = .03$). The benefit of abciximab was confined only to patients with troponin level elevation (13.1% vs. 18.3%, RR 0.71; 95% CI, 0.54–0.95; $P = .02$).[113] The ISAR-REACT 4 trial enrolled patients with troponin-positive ACS and pretreated with clopidogrel 600 mg.[114] However, there were no differences in the primary ischemic endpoint between abciximab and bivalirudin (10.9% vs. 11.0%, RR 0.99; 95% CI, 0.74–1.32; $P = .94$), with abciximab having an increased risk of major bleeding (4.6% vs. 2.6%, RR, 1.84; 95% CI, 1.10–3.07; $P = .02$).

In a meta-analysis of patients with STEMI undergoing primary PCI, abciximab reduced short- (at 30 days, 2.4% vs. 6.2%, $P = .047$) and long-term (at 6–12 months, 4.4% vs. 6.2%, $P = .01$) mortality.[115] However, in the Bavarian Reperfusion Alternatives Evaluation 3 (BRAVE 3) trial, "upstream" (before catheterization laboratory entry) administration of abciximab did not reduce infarct size (15.7% vs. 16.6%, $P = .47$) and the composite of death, MI, or urgent revascularization at 30 days in patients with STEMI undergoing primary PCI (5.0% vs. 3.8%, RR 1.3; 95% CI, 0.7–2.6; $P = .40$).[116] Interestingly, in the Intracoronary Abciximab and Aspiration Thrombectomy in Patients with Large Anterior Myocardial Infarction (INFUSE-AMI) trial, the infarct size at 30 days was significantly reduced by bolus intracoronary abciximab delivered to the infarct lesion by means of a perfusion balloon (15.1% vs. 17.9%, $P = .03$) but not by manual thrombectomy (17.0% vs. 17.3%, $P = .51$).[117]

Eptifibatide

Eptifibatide is a synthetic cyclic Lys-Gly-Asp (KGD)-containing heptapeptide derived from a protein found in the venom of a rattlesnake, making it highly specific for the GP IIb/IIIa receptor.[118,119] With the recommended double-bolus and infusion regimen (180 µg/kg, followed by a second 180 µg/kg bolus, followed by 2 µg/kg for a minimum of 12 h), peak plasma levels are attained within 5 min and a slightly lower concentration is maintained throughout the infusion period.[119–121] Because of its lower affinity for GP IIb/IIIa and relatively rapid decrease in plasma concentration compared with other GP IIb/IIIa inhibitors, the antiplatelet effects of eptifibatide are rapidly reversible. Plasma clearance of eptifibatide occurs with a half-life of 2.5 h, and recovery of platelet aggregation occurs within 4 h of discontinuation of infusion.[119] The majority of the drug is eliminated approximately 50% by renal clearance, and therefore the dose needs to be adjusted in patients with moderate renal dysfunction.

In the Platelet Glycoprotein IIb/IIIa in Unstable Angina: Receptor Suppression Using Integrilin (eptifibatide) Therapy (PURSUIT) trial, eptifibatide reduced the incidence of death or MI compared with placebo in patients with ACS without STEMI at 30 days (14.2% vs. 15.7%, $P = .03$).[121] The Eptifibatide versus Abciximab in primary PCI for Acute Myocardial Infarction (EVA-AMI) trial showed that ST-segment resolution (62.6% vs. 56.3%, $P = .16$) and clinical impact of eptifibatide was similar to abciximab (10.6% vs. 10.9%, $P = .9$).[122] However, in the Early Glycoprotein IIb/IIIa Inhibition in Patients with Non-ST-Segment Elevation Acute Coronary Syndrome (EARLY-ACS) trial, upstream administration of eptifibatide in patients with ACS was associated with an increased risk of bleeding (early vs. delayed eptifibatide, 9.3% vs. 10.0%, OR 1.42; 95% CI, 10.7–1.89; $P = .02$) but there were no significant differences in death or MI at 30 days (early vs. delayed eptifibatide, 11.2% vs. 12.3%, OR 0.89; 95% CI, 0.79–1.01; $P = .08$).[123]

Tirofiban

Tirofiban is a synthetic nonpeptide tyrosine derivative that acts as an Arg-Gly-Asp (RGD) mimetic. Tirofiban has reversible inhibition for the GP IIb/IIIa receptor and a plasma half-life of about 2 h.[124] Its receptor binding avidity is intermediate between that of eptifibatide and abciximab. Although a 10 µg/kg bolus dose inhibited more than 90% platelet aggregation in response to 5 µmol/L ADP, the inhibitory effect of this dose was suboptimal with a higher ADP concentration.[119,125] For this reason, the recommend dose of tirofiban is 25 µg/kg bolus over 3 minutes, followed by 0.15 µg/kg/min infusion for up to 18 h. This high-dose regimen has now replaced the originally approved dosing regimen (0.4 µg/kg/min for 30 min followed by 0.1 µg/kg/min).[96] Platelet function returns to baseline after drug discontinuation within 4–8 h, and tirofiban is mainly cleared by renal mechanism, thus dosage adjustment according to renal function is acquired.[124]

The Additive Value of Tirofiban Administered with the High-dose Bolus in the Prevention of Ischemic Complication during High-risk Coronary Angioplasty (ADVANCE) trial showed a reduction in the composite of death, MI, TVR, and bailout use with tirofiban versus placebo at 6 months (20% vs. 35%, HR 0.51; 95% CI, 0.29–0.88; $P = .01$).[126] In the Ongoing Tirofiban in Myocardial Infarction Evaluation (On-TIME) 2 trial, patients with STEMI were randomly enrolled to high-dose tirofiban or placebo in the prehospital setting.[127] Tirofiban improved ST-segment resolution respectively before (10.9 vs. 12.1 mm, $P = .028$) and after PCI (3.6 vs. 4.8 mm, $P = .003$) without increasing risk of bleeding (4% vs. 3%, $P = .36$) and thrombocytopenia (2.0% vs. 1.8%, $P = .81$). Furthermore, there was significantly less need for thrombotic bailout due to TIMI flow grade 0 to 2 and abrupt closure of the culprit vessel in the tirofiban group (19.9% vs. 28.5%, $P = .002$). However, the rates of the composite of death, recurrent MI, and TVR were not significantly different between tirofiban and placebo (7.0% vs. 8.2%, $P = .485$).

Phosphodiesterase inhibitor

Both cAMP and cyclic guanosine monophosphate (cGMP) are critical intracellular second messengers that regulate cellular signaling and function.[128] Platelet cAMP and cGMP are hydrolyzed to inactive AMP and GMP by phosphodiesterase (PDE) 2 and PDE3, respectively.[129] Increased intracellular levels of cAMP and cGMP in platelets are associated with attenuated platelet aggregation in response to various agonists.[129] Therefore inhibition of PDEs limits the hydrolysis of cyclic nucleotides and increase cAMP and cGMP levels facilitating platelet inhibition.

Cilostazol

Cilostazol is a 2-oxo-quinoline derivative that selectively inhibits PDE3 and the active cellular uptake of adenosine.[130] A dose of 100 mg of cilostazol reaches peak plasma concentration in 3 h and has a half-life of approximately 11 h.[131]

Because cilostazol is metabolized primarily in liver via CYP3A5 or CYP2C19, the dose should be reduced in patients with coadministration of CYP3A5 or CYP2C19 inhibitors.[130] Cilostazol is contraindicated in patients with congestive heart failure because of its increased mortality risk.[129,132]

In patients with type 2 DM who have reduced platelet inhibition, the addition of cilostazol to aspirin and clopidogrel enhances ADP-induced platelet inhibition compared with aspirin and clopidogrel alone.[133] In patients with CAD undergoing PCI, triple antiplatelet therapy (aspirin, clopidogrel, and cilostazol) compared with DAPT (aspirin and clopidogrel) resulted in fewer deaths, MI, or target lesion revascularization (incident rate ratio 0.68; 95% CI, 0.60−0.78) irrespective of the stent type and without increasing the risk of bleeding.[134] However, triple antiplatelet therapy for overcoming high on-treatment platelet reactivity of clopidogrel is rarely used in light of the availability of the more potent $P2Y_{12}$ inhibitors (e.g., prasugrel or ticagrelor). Interestingly, in a meta-analysis of 36 randomized controlled trials involving 82,144 patients, cilostazol alone showed better outcomes than other antiplatelet medications in the long-term secondary prevention of transient ischemic attack or ischemic stroke.[135] Despite these clinical benefits, guidelines recommend cilostazol only for the treatment of intermittent claudication in patients with PAD based on data showing improvement on maximal walking distance, pain-free walking distance, and quality of life.[136]

Additional targets

There are several alternative targets for novel drug development. These include targets modulating platelet (1) adhesion, (2) signaling, and (3) secretion. (1) Caplacizumab (ALX-0081) is an anti-vWF humanized single-variable-domain immunoglobulin that binds to the A1 domain of vWF and inhibits its interaction with GP Ib-IX-V.[137] Revacept is the humanized Fc fusion protein of the GP VI ectodomain that prevents collagen-induced platelet aggregation without affecting general hemostasis.[138] (2) Although the inhibition of intracellular signaling of platelets can affect other cellular mechanisms, the PI3Kβ inhibitor (AZD6482) in combination with aspirin shows a greater antiplatelet effect compared with clopidogrel plus aspirin with significantly less bleeding risk.[139] Among various tyrosine kinases related to signaling through the GP VI, CLEC2, and FcγRIIA receptors, Bruton tyrosine kinase inhibitors significantly suppress GP Ib- and GP VI-dependent thrombus formation with sparing hemostatic platelet function.[140] (3) In animal models, APT 102, recombinant protein of apyrase, substantially enhances scavenging of extracellular ADP thus preventing thrombotic reocclusion and decreasing the infarct size compared with clopidogrel.[141]

References

1. Davi G, Patrono C. Platelet activation and atherothrombosis. *N Engl J Med.* 2007; 357(24):2482−2494.
2. Moon JY, Franchi F, Rollini F, Angiolillo DJ. Evolution of coronary stent technology and implications for duration of dual antiplatelet therapy. *Prog Cardiovasc Dis.* 2018; 60(4−5):478−490.
3. Capodanno D, Mehran R, Valgimigli M, et al. Aspirin-free strategies in cardiovascular disease and cardioembolic stroke prevention. *Nat Rev Cardiol.* 2018;15(8):480−496.
4. Kaushansky K. Historical review: megakaryopoiesis and thrombopoiesis. *Blood.* 2008; 111(3):981−986.
5. Quach ME, Chen W, Li R. Mechanisms of platelet clearance and translation to improve platelet storage. *Blood.* 2018;131(14):1512−1521.
6. Michelson AD, Cattaneo M, Frelinger AL, Newman PJ. *Platelet Structure. Platelets.* 3rd ed. 2013:117−144.
7. Rivera J, Lozano ML, Navarro-Nunez L, Vicente V. Platelet receptors and signaling in the dynamics of thrombus formation. *Haematologica.* 2009;94(5):700−711.
8. Gremmel T, Frelinger 3rd AL, Michelson AD. Platelet physiology. *Semin Thromb Hemost.* 2016;42(3):191−204.
9. Estevez B, Du X. New concepts and mechanisms of platelet activation signaling. *Physiology.* 2017;32(2):162−177.
10. Franchi F, Angiolillo DJ. Novel antiplatelet agents in acute coronary syndrome. *Nat Rev Cardiol.* 2015;12(1):30−47.
11. Versteeg HH, Heemskerk JW, Levi M, Reitsma PH. New fundamentals in hemostasis. *Physiol Rev.* 2013;93(1):327−358.
12. Varga-Szabo D, Pleines I, Nieswandt B. Cell adhesion mechanisms in platelets. *Arterioscler Thromb Vasc Biol.* 2008;28(3):403−412.
13. Offermanns S. Activation of platelet function through G protein-coupled receptors. *Circ Res.* 2006;99(12):1293−1304.
14. Angiolillo DJ, Capodanno D, Goto S. Platelet thrombin receptor antagonism and atherothrombosis. *Eur Heart J.* 2010;31(1):17−28.
15. Kahn ML, Zheng YW, Huang W, et al. A dual thrombin receptor system for platelet activation. *Nature.* 1998;394(6694):690−694.
16. Coughlin SR. Protease-activated receptors in hemostasis, thrombosis and vascular biology. *J Thromb Haemostasis.* 2005;3(8):1800−1814.
17. Arachiche A, Mumaw MM, de la Fuente M, Nieman MT. Protease-activated receptor 1 (PAR1) and PAR4 heterodimers are required for PAR1-enhanced cleavage of PAR4 by alpha-thrombin. *J Biol Chem.* 2013;288(45):32553−32562.
18. De Candia E. Mechanisms of platelet activation by thrombin: a short history. *Thromb Res.* 2012;129(3):250−256.
19. Tricoci P, Huang Z, Held C, et al. Thrombin-receptor antagonist vorapaxar in acute coronary syndromes. *N Engl J Med.* 2012;366(1):20−33.
20. Dorsam RT, Kunapuli SP. Central role of the P2Y12 receptor in platelet activation. *J Clin Invest.* 2004;113(3):340−345.
21. Jin J, Daniel JL, Kunapuli SP. Molecular basis for ADP-induced platelet activation. II. The P2Y1 receptor mediates ADP-induced intracellular calcium mobilization and shape change in platelets. *J Biol Chem.* 1998;273(4):2030−2034.

22. Hardy AR, Jones ML, Mundell SJ, Poole AW. Reciprocal cross-talk between P2Y1 and P2Y12 receptors at the level of calcium signaling in human platelets. *Blood*. 2004; 104(6):1745−1752.

23. Khan A, Li D, Ibrahim S, Smyth E, Woulfe DS. The physical association of the P2Y12 receptor with PAR4 regulates arrestin-mediated Akt activation. *Mol Pharmacol*. 2014; 86(1):1−11.

24. Yusuf S, Zhao F, Mehta SR, Chrolavicius S, Tognoni G, Fox KK. Effects of clopidogrel in addition to aspirin in patients with acute coronary syndromes without ST-segment elevation. *N Engl J Med*. 2001;345(7):494−502.

25. Wiviott SD, Braunwald E, McCabe CH, et al. Prasugrel versus clopidogrel in patients with acute coronary syndromes. *N Engl J Med*. 2007;357(20):2001−2015.

26. Wallentin L, Becker RC, Budaj A, et al. Ticagrelor versus clopidogrel in patients with acute coronary syndromes. *N Engl J Med*. 2009;361(11):1045−1057.

27. Adler DH, Cogan JD, Phillips 3rd JA, et al. Inherited human cPLA(2alpha) deficiency is associated with impaired eicosanoid biosynthesis, small intestinal ulceration, and platelet dysfunction. *J Clin Invest*. 2008;118(6):2121−2131.

28. Patrono C, García Rodríguez LA, Landolfi R, Baigent C. Low-dose aspirin for the prevention of atherothrombosis. *N Engl J Med*. 2005;353(22):2373−2383.

29. FitzGerald GA, Oates JA, Hawiger J, et al. Endogenous biosynthesis of prostacyclin and thromboxane and platelet function during chronic administration of aspirin in man. *J Clin Invest*. 1983;71(3):676−688.

30. Thomas DW, Mannon RB, Mannon PJ, et al. Coagulation defects and altered hemodynamic responses in mice lacking receptors for thromboxane A2. *J Clin Invest*. 1998; 102(11):1994−2001.

31. Moers A, Wettschureck N, Gruner S, Nieswandt B, Offermanns S. Unresponsiveness of platelets lacking both Galpha(q) and Galpha(13). Implications for collagen-induced platelet activation. *J Biol Chem*. 2004;279(44):45354−45359.

32. Plow EF, Byzova T. The biology of glycoprotein IIb-IIIa. *Coron Artery Dis*. 1999;10(8): 547−551.

33. Kulkarni S, Dopheide SM, Yap CL, et al. A revised model of platelet aggregation. *J Clin Invest*. 2000;105(6):783−791.

34. Boersma E, Harrington RA, Moliterno DJ, et al. Platelet glycoprotein IIb/IIIa inhibitors in acute coronary syndromes: a meta-analysis of all major randomised clinical trials. *Lancet*. 2002;359(9302):189−198.

35. Botting RM. Vane's discovery of the mechanism of action of aspirin changed our understanding of its clinical pharmacology. *Pharmacol Rep*. 2010;62(3):518−525.

36. Rowland M, Riegelman S, Harris PA, Sholkoff SD. Absorption kinetics of aspirin in man following oral administration of an aqueous solution. *J Pharmacol Sci*. 1972; 61(3):379−385.

37. Burch JW, Stanford N, Majerus PW. Inhibition of platelet prostaglandin synthetase by oral aspirin. *J Clin Invest*. 1978;61(2):314−319.

38. Gargiulo G, Windecker S, Vranckx P, Gibson CM, Mehran R, Valgimigli M. A critical appraisal of aspirin in secondary prevention: is less more? *Circulation*. 2016;134(23): 1881−1906.

39. Patrignani P, Filabozzi P, Patrono C. Selective cumulative inhibition of platelet thromboxane production by low-dose aspirin in healthy subjects. *J Clin Invest*. 1982;69(6): 1366−1372.

40. Capodanno D, Ingala S, Calderone D, Angiolillo DJ. Aspirin for the primary prevention of cardiovascular disease: latest evidence. *Expert Rev Cardiovasc Ther.* 2019;17(9): 633−643.
41. Baigent C, Blackwell L, Collins R, et al. Aspirin in the primary and secondary prevention of vascular disease: collaborative meta-analysis of individual participant data from randomised trials. *Lancet.* 2009;373(9678):1849−1860.
42. Collaborative meta-analysis of randomised trials of antiplatelet therapy for prevention of death, myocardial infarction, and stroke in high risk patients. *Br Med J.* 2002; 324(7329):71−86.
43. Smith Jr SC, Benjamin EJ, Bonow RO, et al. AHA/ACCF secondary prevention and risk reduction therapy for patients with coronary and other atherosclerotic vascular disease: 2011 update: a guideline from the American Heart Association and American College of Cardiology Foundation. *Circulation.* 2011;124(22):2458−2473.
44. Vandvik PO, Lincoff AM, Gore JM, et al. Primary and secondary prevention of cardio-vascular disease: antithrombotic therapy and prevention of thrombosis, 9th ed: American College of Chest Physicians Evidence-Based Clinical Practice Guidelines. *Chest.* 2012;141(2 Suppl):e637S−e668S.
45. Levine GN, Bates ER, Bittl JA, et al. 2016 ACC/AHA guideline focused update on duration of dual antiplatelet therapy in patients with coronary artery disease: a report of the American College of Cardiology/American Heart Association Task Force on Clinical Practice Guidelines: an update of the 2011 ACCF/AHA/SCAI guideline for percutaneous coronary intervention, 2011 ACCF/AHA guideline for coronary artery bypass graft surgery, 2012 ACC/AHA/ACP/AATS/PCNA/SCAI/STS guideline for the Diagnosis and management of patients with stable ischemic heart disease, 2013 ACCF/AHA guideline for the management of ST-elevation myocardial infarction, 2014 AHA/ACC guideline for the management of patients with non-ST-elevation acute coronary syndromes, and 2014 ACC/AHA guideline on perioperative cardiovascular evaluation and management of patients undergoing noncardiac surgery. *Circulation.* 2016;134(10):e123−155.
46. Mehta SR, Tanguay JF, Eikelboom JW, et al. Double-dose versus standard-dose clopidogrel and high-dose versus low-dose aspirin in individuals undergoing percutaneous coronary intervention for acute coronary syndromes (CURRENT-OASIS 7): a randomised factorial trial. *Lancet.* 2010;376(9748):1233−1243.
47. Xian Y, Wang TY, McCoy LA, et al. Association of discharge aspirin dose with outcomes after acute myocardial infarction: insights from the treatment with ADP receptor inhibitors: longitudinal assessment of treatment Patterns and events after acute coronary syndrome (TRANSLATE-ACS) study. *Circulation.* 2015;132(3):174−181.
48. Johnston A, Jones WS, Hernandez AF. The ADAPTABLE trial and aspirin dosing in secondary prevention for patients with coronary artery disease. *Curr Cardiol Rep.* 2016;18(8):81.
49. Rocca B, Santilli F, Pitocco D, et al. The recovery of platelet cyclooxygenase activity explains interindividual variability in responsiveness to low-dose aspirin in patients with and without diabetes. *J Thromb Haemostasis.* 2012;10(7):1220−1230.
50. Capodanno D, Patel A, Dharmashankar K, et al. Pharmacodynamic effects of different aspirin dosing regimens in type 2 diabetes mellitus patients with coronary artery disease. *Circ Cardiovasc Interv.* 2011;4(2):180−187.
51. Raslan Z, Naseem KM. The control of blood platelets by cAMP signalling. *Biochem Soc Trans.* 2014;42(2):289−294.

52. Yang J, Wu J, Jiang H, et al. Signaling through Gi family members in platelets. Redundancy and specificity in the regulation of adenylyl cyclase and other effectors. *J Biol Chem*. 2002;277(48):46035–46042.

53. Bertrand ME, Rupprecht HJ, Urban P, Gershlick AH. Double-blind study of the safety of clopidogrel with and without a loading dose in combination with aspirin compared with ticlopidine in combination with aspirin after coronary stenting : the clopidogrel aspirin stent international cooperative study (CLASSICS). *Circulation*. 2000;102(6):624–629.

54. Kazui M, Nishiya Y, Ishizuka T, et al. Identification of the human cytochrome P450 enzymes involved in the two oxidative steps in the bioactivation of clopidogrel to its pharmacologically active metabolite. *Drug Metab Dispos*. 2010;38(1):92–99.

55. Bates ER, Lau WC, Angiolillo DJ. Clopidogrel-drug interactions. *J Am Coll Cardiol*. 2011;57(11):1251–1263.

56. von Beckerath N, Taubert D, Pogatsa-Murray G, Schomig E, Kastrati A, Schomig A. Absorption, metabolization, and antiplatelet effects of 300-, 600-, and 900-mg loading doses of clopidogrel: results of the ISAR-CHOICE (intracoronary stenting and antithrombotic regimen: choose between 3 high oral doses for immediate clopidogrel effect) trial. *Circulation*. 2005;112(19):2946–2950.

57. Angiolillo DJ, Bernardo E, Palazuelos J, et al. Functional impact of high clopidogrel maintenance dosing in patients undergoing elective percutaneous coronary interventions. Results of a randomized study. *Thromb Haemostasis*. 2008;99(1):161–168.

58. A randomised, blinded, trial of clopidogrel versus aspirin in patients at risk of ischaemic events (CAPRIE). CAPRIE Steering Committee. *Lancet*. 1996;348(9038):1329–1339.

59. Chen ZM, Jiang LX, Chen YP, et al. Addition of clopidogrel to aspirin in 45,852 patients with acute myocardial infarction: randomised placebo-controlled trial. *Lancet*. 2005;366(9497):1607–1621.

60. Sabatine MS, Cannon CP, Gibson CM, et al. Addition of clopidogrel to aspirin and fibrinolytic therapy for myocardial infarction with ST-segment elevation. *N Engl J Med*. 2005;352(12):1179–1189.

61. Steinhubl SR, Berger PB, Mann 3rd JT, et al. Early and sustained dual oral antiplatelet therapy following percutaneous coronary intervention: a randomized controlled trial. *J Am Med Assoc*. 2002;288(19):2411–2420.

62. Bhatt DL, Fox KA, Hacke W, et al. Clopidogrel and aspirin versus aspirin alone for the prevention of atherothrombotic events. *N Engl J Med*. 2006;354(16):1706–1717.

63. Angiolillo DJ, Fernandez-Ortiz A, Bernardo E, et al. Variability in individual responsiveness to clopidogrel: clinical implications, management, and future perspectives. *J Am Coll Cardiol*. 2007;49(14):1505–1516.

64. Angiolillo DJ, Fernandez-Ortiz A, Bernardo E, et al. High clopidogrel loading dose during coronary stenting: effects on drug response and interindividual variability. *Eur Heart J*. 2004;25(21):1903–1910.

65. Sibbing D, Aradi D, Alexopoulos D, et al. Updated expert consensus statement on platelet function and genetic testing for guiding P2Y12 receptor inhibitor treatment in percutaneous coronary intervention. *JACC Cardiovasc Interv*. 2019;12(16):1521–1537.

66. Angiolillo DJ, Gibson CM, Cheng S, et al. Differential effects of omeprazole and pantoprazole on the pharmacodynamics and pharmacokinetics of clopidogrel in healthy subjects: randomized, placebo-controlled, crossover comparison studies. *Clin Pharmacol Ther*. 2011;89(1):65–74.

67. Sugidachi A, Asai F, Ogawa T, Inoue T, Koike H. The in vivo pharmacological profile of CS-747, a novel antiplatelet agent with platelet ADP receptor antagonist properties. *Br J Pharmacol*. 2000;129(7):1439–1446.

68. Hagihara K, Kazui M, Kurihara A, et al. Biotransformation of prasugrel, a novel thienopyridine antiplatelet agent, to the pharmacologically active metabolite. *Drug Metab Dispos*. 2010;38(6):898–904.

69. Niitsu Y, Jakubowski JA, Sugidachi A, Asai F. Pharmacology of CS-747 (prasugrel, LY640315), a novel, potent antiplatelet agent with in vivo P2Y12 receptor antagonist activity. *Semin Thromb Hemost*. 2005;31(2):184–194.

70. Sugidachi A, Ogawa T, Kurihara A, et al. The greater in vivo antiplatelet effects of prasugrel as compared to clopidogrel reflect more efficient generation of its active metabolite with similar antiplatelet activity to that of clopidogrel's active metabolite. *J Thromb Haemostasis*. 2007;5(7):1545–1551.

71. Jernberg T, Payne CD, Winters KJ, et al. Prasugrel achieves greater inhibition of platelet aggregation and a lower rate of non-responders compared with clopidogrel in aspirin-treated patients with stable coronary artery disease. *Eur Heart J*. 2006;27(10):1166–1173.

72. Roe MT, Armstrong PW, Fox KA, et al. Prasugrel versus clopidogrel for acute coronary syndromes without revascularization. *N Engl J Med*. 2012;367(14):1297–1309.

73. Jakubowski JA, Erlinge D, Alexopoulos D, et al. The rationale for and clinical pharmacology of prasugrel 5 mg. *Am J Cardiovasc Drugs*. 2017;17(2):109–121.

74. Montalescot G, Bolognese L, Dudek D, et al. Pretreatment with prasugrel in non-ST-segment elevation acute coronary syndromes. *N Engl J Med*. 2013;369(11):999–1010.

75. Hoffmann K, Lutz DA, Strassburger J, Baqi Y, Muller CE, von Kugelgen I. Competitive mode and site of interaction of ticagrelor at the human platelet P2Y12 -receptor. *J Thromb Haemostasis*. 2014;12(11):1898–1905.

76. Garcia C, Maurel-Ribes A, Nauze M, et al. Deciphering biased inverse agonism of cangrelor and ticagrelor at P2Y12 receptor. *Cell Mol Life Sci*. 2019;76(3):561–576.

77. Aungraheeta R, Conibear A, Butler M, et al. Inverse agonism at the P2Y12 receptor and ENT1 transporter blockade contribute to platelet inhibition by ticagrelor. *Blood*. 2016;128(23):2717–2728.

78. Teng R, Oliver S, Hayes MA, Butler K. Absorption, distribution, metabolism, and excretion of ticagrelor in healthy subjects. *Drug Metab Dispos*. 2010;38(9):1514–1521.

79. Tantry US, Bliden KP, Wei C, et al. First analysis of the relation between CYP2C19 genotype and pharmacodynamics in patients treated with ticagrelor versus clopidogrel: the ONSET/OFFSET and RESPOND genotype studies. *Circ Cardiovasc Genet*. 2010;3(6):556–566.

80. Bonaca MP, Bhatt DL, Cohen M, et al. Long-term use of ticagrelor in patients with prior myocardial infarction. *N Engl J Med*. 2015;372(19):1791–1800.

81. Schupke S, Neumann FJ, Menichelli M, et al. Ticagrelor or prasugrel in patients with acute coronary syndromes. *N Engl J Med*. 2019;381(16):1524–1534.

82. Steg PG, Bhatt DL, Simon T, et al. Ticagrelor in patients with stable coronary disease and diabetes. *N Engl J Med*. 2019;381(14):1309–1320.

83. Vranckx P, Valgimigli M, Juni P, et al. Ticagrelor plus aspirin for 1 month, followed by ticagrelor monotherapy for 23 months vs aspirin plus clopidogrel or ticagrelor for 12 months, followed by aspirin monotherapy for 12 months after implantation of a drug-eluting stent: a multicentre, open-label, randomised superiority trial. *Lancet*. 2018;392(10151):940–949.

84. Hahn JY, Song YB, Oh JH, et al. Effect of P2Y12 inhibitor monotherapy vs dual antiplatelet therapy on cardiovascular events in patients undergoing percutaneous coronary intervention: the SMART-CHOICE randomized clinical trial. *J Am Med Assoc.* 2019; 321(24):2428−2437.

85. Watanabe H, Domei T, Morimoto T, et al. Effect of 1-month dual antiplatelet therapy followed by clopidogrel vs 12-month dual antiplatelet therapy on cardiovascular and bleeding events in patients receiving PCI: the STOPDAPT-2 randomized clinical trial. *J Am Med Assoc.* 2019;321(24):2414−2427.

86. Mehran R, Baber U, Sharma SK, et al. Ticagrelor with or without aspirin in high-risk patients after PCI. *N Engl J Med.* 2019;381(21):2032−2042.

87. Zanchin T, Temperli F, Karagiannis A, et al. Frequency, reasons, and impact of premature ticagrelor discontinuation in patients undergoing coronary revascularization in routine clinical practice: results from the bern percutaneous coronary intervention registry. *Circ Cardiovasc Interv.* 2018;11(5):e006132.

88. Storey RF, Becker RC, Harrington RA, et al. Pulmonary function in patients with acute coronary syndrome treated with ticagrelor or clopidogrel (from the Platelet Inhibition and Patient Outcomes [PLATO] pulmonary function substudy). *Am J Cardiol.* 2011; 108(11):1542−1546.

89. Ingall AH, Dixon J, Bailey A, et al. Antagonists of the platelet P2T receptor: a novel approach to antithrombotic therapy. *J Med Chem.* 1999;42(2):213−220.

90. Akers WS, Oh JJ, Oestreich JH, Ferraris S, Wethington M, Steinhubl SR. Pharmacokinetics and pharmacodynamics of a bolus and infusion of cangrelor: a direct, parenteral P2Y12 receptor antagonist. *J Clin Pharmacol.* 2010;50(1):27−35.

91. Angiolillo DJ, Schneider DJ, Bhatt DL, et al. Pharmacodynamic effects of cangrelor and clopidogrel: the platelet function substudy from the cangrelor versus standard therapy to achieve optimal management of platelet inhibition (CHAMPION) trials. *J Thromb Thrombolysis.* 2012;34(1):44−55.

92. Schneider DJ, Seecheran N, Raza SS, Keating FK, Gogo P. Pharmacodynamic effects during the transition between cangrelor and prasugrel. *Coron Artery Dis.* 2015;26(1): 42−48.

93. Angiolillo DJ, Rollini F, Storey RF, et al. International expert consensus on switching platelet P2Y12 receptor-inhibiting therapies. *Circulation.* 2017;136(20):1955−1975.

94. Schneider DJ, Agarwal Z, Seecheran N, Keating FK, Gogo P. Pharmacodynamic effects during the transition between cangrelor and ticagrelor. *JACC Cardiovasc Interv.* 2014; 7(4):435−442.

95. Bhatt DL, Stone GW, Mahaffey KW, et al. Effect of platelet inhibition with cangrelor during PCI on ischemic events. *N Engl J Med.* 2013;368(14):1303−1313.

96. Capodanno D, Milluzzo RP, Angiolillo DJ. Intravenous antiplatelet therapies (glycoprotein IIb/IIIa receptor inhibitors and cangrelor) in percutaneous coronary intervention: from pharmacology to indications for clinical use. *Ther Adv Cardiovasc Dis.* 2019; 13, 1753944719893274.

97. Franchi F, Rollini F, Rivas A, et al. Platelet inhibition with cangrelor and crushed ticagrelor in patients with ST-segment-elevation myocardial infarction undergoing primary percutaneous coronary intervention. *Circulation.* 2019;139(14):1661−1670.

98. Angiolillo DJ, Firstenberg MS, Price MJ, et al. Bridging antiplatelet therapy with cangrelor in patients undergoing cardiac surgery: a randomized controlled trial. *J Am Med Assoc.* 2012;307(3):265−274.

99. Vu TK, Hung DT, Wheaton VI, Coughlin SR. Molecular cloning of a functional thrombin receptor reveals a novel proteolytic mechanism of receptor activation. *Cell*. 1991;64(6):1057−1068.

100. Chackalamannil S, Wang Y, Greenlee WJ, et al. Discovery of a novel, orally active himbacine-based thrombin receptor antagonist (SCH 530348) with potent antiplatelet activity. *J Med Chem*. 2008;51(11):3061−3064.

101. Ghosal A, Lu X, Penner N, et al. Identification of human liver cytochrome P450 enzymes involved in the metabolism of SCH 530348 (Vorapaxar), a potent oral thrombin protease-activated receptor 1 antagonist. *Drug Metab Dispos*. 2011;39(1):30−38.

102. Kosoglou T, Reyderman L, Tiessen RG, et al. Pharmacodynamics and pharmacokinetics of the novel PAR-1 antagonist vorapaxar (formerly SCH 530348) in healthy subjects. *Eur J Clin Pharmacol*. 2012;68(3):249−258.

103. Becker RC, Moliterno DJ, Jennings LK, et al. Safety and tolerability of SCH 530348 in patients undergoing non-urgent percutaneous coronary intervention: a randomised, double-blind, placebo-controlled phase II study. *Lancet*. 2009;373(9667):919−928.

104. Kato Y, Kita Y, Hirasawa-Taniyama Y, et al. Inhibition of arterial thrombosis by a protease-activated receptor 1 antagonist, FR171113, in the Guinea pig. *Eur J Pharmacol*. 2003;473(2−3):163−169.

105. Morrow DA, Braunwald E, Bonaca MP, et al. Vorapaxar in the secondary prevention of atherothrombotic events. *N Engl J Med*. 2012;366(15):1404−1413.

106. Scirica BM, Bonaca MP, Braunwald E, et al. Vorapaxar for secondary prevention of thrombotic events for patients with previous myocardial infarction: a prespecified subgroup analysis of the TRA 2 degrees P TIMI 50 trial. *Lancet*. 2012;380(9850): 1317−1324.

107. Bonaca MP, Scirica BM, Creager MA, et al. Vorapaxar in patients with peripheral artery disease: results from TRA2{degrees}P-TIMI 50. *Circulation*. 2013;127(14), 1522−1529, 1529e1521−1526.

108. Moon JY, Franchi F, Rollini F, Angiolillo DJ. Role for thrombin receptor antagonism with vorapaxar in secondary prevention of atherothrombotic events: from bench to bedside. *J Cardiovasc Pharmacol Therapeut*. 2018;23(1):23−37.

109. Coller BS. A new murine monoclonal antibody reports an activation-dependent change in the conformation and/or microenvironment of the platelet glycoprotein IIb/IIIa complex. *J Clin Invest*. 1985;76(1):101−108.

110. Tcheng JE, Ellis SG, George BS, et al. Pharmacodynamics of chimeric glycoprotein IIb/IIIa integrin antiplatelet antibody Fab 7E3 in high-risk coronary angioplasty. *Circulation*. 1994;90(4):1757−1764.

111. Mascelli MA, Lance ET, Damaraju L, Wagner CL, Weisman HF, Jordan RE. Pharmacodynamic profile of short-term abciximab treatment demonstrates prolonged platelet inhibition with gradual recovery from GP IIb/IIIa receptor blockade. *Circulation*. 1998;97(17):1680−1688.

112. Kastrati A, Mehilli J, Schuhlen H, et al. A clinical trial of abciximab in elective percutaneous coronary intervention after pretreatment with clopidogrel. *N Engl J Med*. 2004; 350(3):232−238.

113. Kastrati A, Mehilli J, Neumann FJ, et al. Abciximab in patients with acute coronary syndromes undergoing percutaneous coronary intervention after clopidogrel pretreatment: the ISAR-REACT 2 randomized trial. *J Am Med Assoc*. 2006;295(13): 1531−1538.

114. Kastrati A, Neumann FJ, Schulz S, et al. Abciximab and heparin versus bivalirudin for non-ST-elevation myocardial infarction. *N Engl J Med*. 2011;365(21):1980−1989.
115. De Luca G, Suryapranata H, Stone GW, et al. Abciximab as adjunctive therapy to reperfusion in acute ST-segment elevation myocardial infarction: a meta-analysis of randomized trials. *J Am Med Assoc*. 2005;293(14):1759−1765.
116. Mehilli J, Kastrati A, Schulz S, et al. Abciximab in patients with acute ST-segment-elevation myocardial infarction undergoing primary percutaneous coronary intervention after clopidogrel loading: a randomized double-blind trial. *Circulation*. 2009;119(14):1933−1940.
117. Stone GW, Maehara A, Witzenbichler B, et al. Intracoronary abciximab and aspiration thrombectomy in patients with large anterior myocardial infarction: the INFUSE-AMI randomized trial. *J Am Med Assoc*. 2012;307(17):1817−1826.
118. Scarborough RM, Rose JW, Naughton MA, et al. Characterization of the integrin specificities of disintegrins isolated from American pit viper venoms. *J Biol Chem*. 1993;268(2):1058−1065.
119. Kleiman NS. Pharmacokinetics and pharmacodynamics of glycoprotein IIb-IIIa inhibitors. *Am Heart J*. 1999;138(4 Pt 2):263−275.
120. Phillips DR, Teng W, Arfsten A, et al. Effect of Ca^{2+} on GP IIb-IIIa interactions with integrilin: enhanced GP IIb-IIIa binding and inhibition of platelet aggregation by reductions in the concentration of ionized calcium in plasma anticoagulated with citrate. *Circulation*. 1997;96(5):1488−1494.
121. Inhibition of platelet glycoprotein IIb/IIIa with eptifibatide in patients with acute coronary syndromes. *N Engl J Med*. 1998;339(7):436−443.
122. Zeymer U, Margenet A, Haude M, et al. Randomized comparison of eptifibatide versus abciximab in primary percutaneous coronary intervention in patients with acute ST-segment elevation myocardial infarction: results of the EVA-AMI trial. *J Am Coll Cardiol*. 2010;56(6):463−469.
123. Giugliano RP, White JA, Bode C, et al. Early versus delayed, provisional eptifibatide in acute coronary syndromes. *N Engl J Med*. 2009;360(21):2176−2190.
124. Barrett JS, Murphy G, Peerlinck K, et al. Pharmacokinetics and pharmacodynamics of MK-383, a selective non-peptide platelet glycoprotein-IIb/IIIa receptor antagonist, in healthy men. *Clin Pharmacol Ther*. 1994;56(4):377−388.
125. Kereiakes DJ, Kleiman NS, Ambrose J, et al. Randomized, double-blind, placebo-controlled dose-ranging study of tirofiban (MK-383) platelet IIb/IIIa blockade in high risk patients undergoing coronary angioplasty. *J Am Coll Cardiol*. 1996;27(3):536−542.
126. Valgimigli M, Percoco G, Barbieri D, et al. The additive value of tirofiban administered with the high-dose bolus in the prevention of ischemic complications during high-risk coronary angioplasty: the ADVANCE trial. *J Am Coll Cardiol*. 2004;44(1):14−19.
127. Van't Hof AW, Ten Berg J, Heestermans T, et al. Prehospital initiation of tirofiban in patients with ST-elevation myocardial infarction undergoing primary angioplasty (On-TIME 2): a multicentre, double-blind, randomised controlled trial. *Lancet*. 2008;372(9638):537−546.
128. Haslam RJ, Dickinson NT, Jang EK. Cyclic nucleotides and phosphodiesterases in platelets. *Thromb Haemostasis*. 1999;82(2):412−423.
129. Rondina MT, Weyrich AS. Targeting phosphodiesterases in anti-platelet therapy. *Handb Exp Pharmacol*. 2012;210:225−238.

130. Schror K. The pharmacology of cilostazol. *Diabetes Obes Metabol.* 2002;4(Suppl 2): S14—S19.
131. Yamamoto H, Takahashi K, Watanabe H, et al. Evaluation of the antiplatelet effects of cilostazol, a phosphodiesterase 3 inhibitor, by VASP phosphorylation and platelet aggregation. *Circ J.* 2008;72(11):1844—1851.
132. Cone J, Wang S, Tandon N, et al. Comparison of the effects of cilostazol and milrinone on intracellular cAMP levels and cellular function in platelets and cardiac cells. *J Cardiovasc Pharmacol.* 1999;34(4):497—504.
133. Angiolillo DJ, Capranzano P, Goto S, et al. A randomized study assessing the impact of cilostazol on platelet function profiles in patients with diabetes mellitus and coronary artery disease on dual antiplatelet therapy: results of the OPTIMUS-2 study. *Eur Heart J.* 2008;29(18):2202—2211.
134. Bangalore S, Singh A, Toklu B, et al. Efficacy of cilostazol on platelet reactivity and cardiovascular outcomes in patients undergoing percutaneous coronary intervention: insights from a meta-analysis of randomised trials. *Open Heart.* 2014;1(1):e000068.
135. Niu PP, Guo ZN, Jin H, Xing YQ, Yang Y. Antiplatelet regimens in the long-term secondary prevention of transient ischaemic attack and ischaemic stroke: an updated network meta-analysis. *BMJ Open.* 2016;6(3):e009013.
136. Thompson PD, Zimet R, Forbes WP, Zhang P. Meta-analysis of results from eight randomized, placebo-controlled trials on the effect of cilostazol on patients with intermittent claudication. *Am J Cardiol.* 2002;90(12):1314—1319.
137. Bartunek J, Barbato E, Heyndrickx G, Vanderheyden M, Wijns W, Holz JB. Novel antiplatelet agents: ALX-0081, a Nanobody directed towards von Willebrand factor. *J Cardiovasc Transl Res.* 2013;6(3):355—363.
138. Ungerer M, Rosport K, Bultmann A, et al. Novel antiplatelet drug revacept (Dimeric Glycoprotein VI-Fc) specifically and efficiently inhibited collagen-induced platelet aggregation without affecting general hemostasis in humans. *Circulation.* 2011;123(17): 1891—1899.
139. Nylander S, Wagberg F, Andersson M, Skarby T, Gustafsson D. Exploration of efficacy and bleeding with combined phosphoinositide 3-kinase beta inhibition and aspirin in man. *J Thromb Haemostasis.* 2015;13(8):1494—1502.
140. Busygina K, Jamasbi J, Seiler T, et al. Oral Bruton tyrosine kinase inhibitors selectively block atherosclerotic plaque-triggered thrombus formation in humans. *Blood.* 2018; 131(24):2605—2616.
141. Moeckel D, Jeong SS, Sun X, et al. Optimizing human apyrase to treat arterial thrombosis and limit reperfusion injury without increasing bleeding risk. *Sci Transl Med.* 2014;6(248):248ra105.

The development of dual antiplatelet therapy: physiologic and clinical implications of multiple pathway inhibition of platelet function

Adam T. Phillips, MD, C. Michael Gibson, MD

Division of Cardiovascular Medicine, Department of Medicine, Beth Israel Deaconess Medical Center, Harvard Medical School, Boston, MA, United States; Baim Institute for Clinical Research, Boston, MA, United States

Introduction

Dual antiplatelet therapy (DAPT) has become an integral part of the care of patients with vascular disease. It evolved in conjunction with percutaneous coronary interventions (PCIs) for acute coronary syndromes (ACSs) but has also been evaluated for patients with chronic coronary syndrome and peripheral arterial disease (PAD). This chapter will focus on the evidence generation for DAPT for patients with vascular disease (Table 3.1).

Aspirin monotherapy

Aspirin irreversibly inhibits cyclooxygenase-dependent platelet aggregation. Because aspirin was the first in the antiplatelet class and is inexpensive, it forms the foundation of antiplatelet strategies. Long-term randomized studies of aspirin monotherapy for secondary prevention in patients with a history of ACS found a 25% risk reduction in major adverse cardiovascular events (reinfarction, stroke, and vascular death).[1]

Its apparent benefit in long-term secondary prevention led to a large study in the ACS population in ISIS-2 (Second International Study of Infarct Survival) ST-elevation myocardial infarction (STEMI) study, which randomized 17,187 patients within 24 h of ACS to intravenous streptokinase, oral aspirin, both, or neither and examined the outcome of vascular death at 5 weeks.[2] The study found a significant reduction in vascular death when aspirin versus placebo was added to streptokinase

Dual Antiplatelet Therapy for Coronary and Peripheral Arterial Disease
https://doi.org/10.1016/B978-0-12-820536-5.00011-2

Table 3.1 Key trials for dual pathway platelet inhibition.

Year	Antiplatelet	Trial	Population	Patients randomized	Key findings
1988	Aspirin	ISIS-2	ACS (suspected)	17,187	Streptokinase plus aspirin reduced vascular mortality at 5 weeks by 42% better than either alone (8% vs. 13.2%)
1998	Ticlopidine	STARS	PCI (successful)	1653	Aspirin plus ticlopidine reduced stent thrombosis compared to aspirin alone or combination of aspirin plus warfarin (0.5% vs. 3.6% vs. 2.7%, $p < .001$)
1996	Clopidogrel	CAPRIE	Atherosclerotic disease (recent ACS, ischemic stroke, or symptomatic PAD)	19,185	Clopidogrel plus aspirin reduced ischemic stroke, MI, or vascular death compared to aspirin alone (5.32% vs. 5.83%, $p = .043$)
2001	Clopidogrel	CURE	ACS (without ST-segment elevation)	12,562	Clopidogrel plus aspirin reduced CV death, nonfatal MI, or stroke compared to aspirin alone (9.3% vs. 11.4%, $p < .001$)
2005	Prasugrel	JUMBO-TIMI 26	PCI (elective or urgent)	904	Prasugrel and clopidogrel result in similarly low bleeding rate (1.7% vs. 1.2%, $p = .59$)
2007	Prasugrel	TRITON-TIMI 38	ACS (moderate-to-high risk with scheduled PCI)	13,608	Prasugrel reduced CV death, nonfatal MI, or nonfatal stroke compared to clopidogrel (9.9% vs. 12.1%, $p < .001$) at the expense of increased life-threatening bleeding events (1.4% vs. 0.9, $p = .01$)
2009	Ticagrelor	PLATO	ACS (with or without ST-segment elevation)	18,624	Ticagrelor reduced CV death, nonfatal MI, or stroke compared to clopidogrel (9.8% vs. 11.7%, $p < .001$) as well as overall mortality (4.5% vs. 5.9%, $p < .001$) at the expense of increased major bleeding (4.5% vs. 3.8%, $p = .03$)
2015	Ticagrelor	PEGASUS-TIMI 54	CAD (ACS 1–3 years prior to enrolment)	21,162	Long-term, low-dose ticagrelor reduced CV death, MI, or stroke compared to aspirin monotherapy (7.7% vs. 9.0%, $p = .004$)
2019	Ticagrelor	TWILIGHT	PCI (3 months prior; moderate-to-high risk of ischemia/bleeding)	7119	3 months of DAPT followed by ticagrelor monotherapy reduced major bleeding (4.0% vs. 7.1%, $p < .001$) without change in death, nonfatal MI, or nonfatal stroke

ACS, acute coronary syndrome; CAD, coronary artery disease; CV, cardiovascular; DAPT, dual antiplatelet therapy; MI, myocardial infarction; PAD, peripheral arterial disease; PCI, percutaneous coronary intervention.

(9.2% vs. 12%) as well as with aspirin alone versus placebo (9.4% vs. 11.8%), and the combination of streptokinase plus aspirin was significantly better than either alone (8% vs. 13.2%, $p < .0001$). This benefit of combination reduced reinfarction (1.8% vs. 2.9%), strokes (0.6% vs. 1.1%), and deaths (8% vs. 13.2%) compared to those allocated to neither therapy. The benefit was durable at 15 months follow up of the study. Streptokinase infusion was associated with increased bleeding requiring transfusion (0.5% vs. 0.2%) and aspirin was not associated with any increase in cerebral hemorrhage or bleeding requiring transfusion.

The results of the ISIS-2 STEMI study set the foundation for antiplatelet therapy to be a cornerstone of management across all ACSs. They also highlighted the balance between ischemic and bleeding events that would direct clinical trials in the antiplatelet space for decades to come. In searching for the most safe and effective balance, dual pathway platelet inhibition was explored with thienopyridines.

Thienopyridines

The thienopyridines, i.e., ticlopidine, clopidogrel, and prasugrel, inhibit ADP-induced platelet aggregation by selectively blocking the P2Y12 receptor.[3] Medications in this class have progressively become better tolerated with fewer side effects.

Ticlopidine

Ticlopidine was the first thienopyridine investigated for the management of percutaneous coronary intervention. The STARS (Stent Anticoagulation Restenosis Study) Investigators randomized 1653 patients receiving a coronary stent to one of three antithrombotic regimens: aspirin alone, aspirin plus warfarin, or aspirin plus ticlopidine.[4] The primary endpoint was a composite of death, revascularization of the target lesion, angiographically evident thrombosis, or myocardial infarction (MI) within 30 days. The investigators found a reduction in the primary endpoint in the aspirin plus ticlopidine group compared to those who received aspirin alone or aspirin plus warfarin (0.5% vs. 3.6% vs. 2.7%, respectively, $p < .001$ for comparison of all three groups). This difference was driven by a reduction in target vessel revascularization (aspirin plus ticlopidine 0.5%, aspirin alone 3.4%, aspirin plus warfarin 2.5%, respectively, $p = .002$) and angiographically evident thrombosis (aspirin plus ticlopidine 0.5%, aspirin alone 2.9%, aspirin plus warfarin 2.7%, respectively, $p = .005$). The rate of bleeding was higher in patients receiving aspirin plus ticlopidine than that in those receiving aspirin alone but lower than that in patients receiving aspirin plus warfarin (5.5% vs. 1.8% vs. 6.2% respectively, $p < .001$ for comparison of all three groups). Interestingly, there were no significant hematologic differences (incidence of neutropenia or thrombocytopenia) among the three groups (aspirin plus ticlopidine 0.5%, aspirin alone 0.2%, aspirin plus warfarin 0.2%, $p = .46$). The STARS suggested that dual pathway platelet inhibition with aspirin and a P2Y12 inhibitor—in this study, it was ticlopidine—was superior to aspirin alone or a combination of aspirin and warfarin to prevent stent thrombosis after stent placement.

Ticlopidine, however, was plagued by its hematologic side effects. Based on clinical trial and registry data from its long-term use in patients receiving the medication for secondary prevention of stroke or coronary disease, the incidence of severe neutropenia and thrombocytopenia was 0.8%.[5] Thrombotic thrombocytopenic purpura was also a rare but observed serious complication.[6] Because of its problematic safety profile, ticlopidine use declined in favor of other, safer, thienopyridines.

Clopidogrel

Clopidogrel, like ticlopidine, is a prodrug that requires activation by the cytochrome p450 system and the active metabolite binds irreversibly to P2Y12. Clopidogrel is an improved drug compared with ticlopidine because it does not have the hematologic side effect profile of neutropenia and thrombocytopenia.

In the CAPRIE study (a randomized, blinded, trial of clopidogrel vs. aspirin in patients at risk of ischemic events), investigators randomized 19,185 patients with atherosclerotic vascular disease (recent ischemic stroke, recent ACS, or symptomatic PAD) to clopidogrel (75 mg daily) or aspirin (325 mg daily) and followed up them for 1−3 years.[7] The primary outcomes was a composite of ischemic stroke, MI, or vascular death. Clopidogrel reduced the risk of the primary outcome compared with aspirin (5.32% vs. 5.83%, $p = .043$). Clopidogrel and aspirin had a similar safety profile, with no increased incidence of bleeding or neutropenia in the clopidogrel group. This was the first study to show that clopidogrel was more effective than aspirin in reducing ischemic risk for patients with ischemic cardiovascular disease and had a similar safety profile to aspirin.

Because clopidogrel and aspirin target different platelet aggregation pathways, the question arose as to whether the combination of aspirin and clopidogrel would be better than either alone. The CURE study (Clopidogrel in Unstable Angina to Prevent Recurrent Events) examined this question. Investigators randomized 12,562 patients with ACS without ST-segment elevation to receive clopidogrel (300 mg loading dose following by 75 mg daily) or placebo added to background aspirin therapy for 3−12 months.[8] The primary endpoint, i.e., cardiovascular death, nonfatal MI, or stroke, occurred less frequently in the clopidogrel group than in the placebo group (9.3% vs. 11.4%, $p < .001$). More patients in the clopidogrel group experienced major bleeding compared with the placebo group (3.7% vs. 2.7%, $p = .001$), but there was no difference in life-threatening bleeding events (2.1% vs. 1.8%, $p = .13$). This study showed that adding clopidogrel to aspirin was better than aspirin alone in patients with ACS without ST-segment elevation. The trial did not test whether adding aspirin to clopidogrel (the superior agent) would improve outcomes. This benefit in reduction of ischemic events was achieved at the expense of increased major bleeding.

The addition of clopidogrel to aspirin appeared to represent a safe and effective method to achieve dual pathway platelet inhibition in patients with coronary disease.

Its ability to inhibit platelet aggregation, however, was variable among patients because of genetic polymorphisms in the cytochrome p450 system facilitating the activation of clopidogrel in the liver. Patients with a loss-of-function mutation in CYP2C19*2, which represents roughly 30% of the population, are functionally resistant to the antiplatelet effects of clopidogrel.[9]

Prasugrel

Prasugrel also requires metabolic activation but exhibits more prompt, potent, predictable platelet inhibition than clopidogrel.[10] Prasugrel was compared to clopidogrel in the phase 2 JUMBO-TIMI 26 (Joint Utilization of Medications to Block Platelets Optimally-Thrombolysis in Myocardial Infarction) study, which randomized 904 patients undergoing PCI and on background aspirin therapy to clopidogrel or prasugrel, followed up them for 30 days, and examined the primary endpoint of clinically significant non-CABG (coronary artery bypass graft)-related bleeding events.[11] Investigators found no difference in the rate of bleeding between prasugrel and clopidogrel (1.7% vs. 1.2%, $p = .59$). This study laid the groundwork for a phase 3 evaluation of prasugrel.

TRITON-TIMI 38 (Trial to Assess Improvement in Therapeutic Outcomes by Optimizing Platelet Inhibition with Prasugrel-Thrombolysis in Myocardial Infarction) randomized 13,608 patients with moderate-to-high risk ACSs with scheduled PCI to receive either prasugrel (60 mg loading dose followed by 10 mg daily dose) or clopidogrel (300 mg loading dose followed by 75 mg daily dose) for 6−15 months and evaluated the primary endpoint of cardiovascular death, nonfatal MI, or nonfatal stroke.[12] Investigators found a reduction in the composite endpoint with prasugrel compared to clopidogrel (9.9% vs. 12.1%, $p < .001$) as well as a reduction in the rate of MI (7.4% vs. 9.7%, $p < .001$), urgent target vessel revascularization (2.5% vs. 3.7%, $p < .001$), and stent thrombosis (1.1% vs. 2.4% vs., $p < .001$). However, patients treated with prasugrel had an increase in major (2.4% vs. 1.8%, $p = .03$) and life-threatening (1.4% vs. 0.9%, $p = .01$) bleeding. Similar to prior studies, improvement in ischemic events appeared to be achieved at the cost of increased major bleeding (including fatal bleeding), and there was no overall mortality difference between the treatment groups. There was not only more bleeding among patients with a prior history of stroke, those weighing <60 kg, and those >75 years of age but also a significant interaction term indicating there was similar or worse efficacy in these subgroups of patients p = .008). Prasugrel is therefore not recommended for these groups of patients or should be used with caution in these patients.

Dipyridamole

Dipyridamole inhibits platelet activation and aggregation by blocking reuptake of adenosine and inhibiting phosphodiesterase-mediated cyclic adenosine monophosphate (cAMP) degradation. While it is a relatively weak antiplatelet agent, when combined with aspirin it can be used for secondary stroke prevention. This combination must be used with caution in patients with coronary artery disease, however, because of the vasodilatory effects of dipyridamole, which can lead to coronary steal and hypotension.

The ESPRIT (Aspirin Plus Dipyridamole vs. Aspirin Alone After Cerebral Ischemia of Artery Origin) study randomized 2739 patients who had a transient ischemic attack or minor stroke from arterial origin within the previous 6 months to either aspirin or aspirin plus dipyridamole.[13] The composite primary endpoint of death from all vascular causes, nonfatal stroke, nonfatal MI, or major bleeding complication was lower in the combination therapy group than that in the aspirin monotherapy group (13% vs. 16%, hazard ratio [HR] 0.80; 95% confidence interval [CI], 0.66−0.98; absolute risk reduction, 1.0% per year). The rate of drug discontinuation, however, was higher in the combination therapy group (470 vs. 184 patients), with most discontinuations due to headache.

The combination of aspirin and dipyridamole was compared to clopidogrel in a study of secondary stroke prevention in 20,332 patients.[14] The primary endpoint of stroke recurrence occurred in 9% of patients receiving combination therapy and in 8.8% of patients receiving clopidogrel (HR 1.01; 95% CI, 0.92−1.11). The rate of hemorrhagic complications was higher in the combination therapy group (4.1% vs. 3.6%; HR 1.15; CI, 1.00−1.32) and the net risk of recurrent stroke or major hemorrhagic event was similar in the two groups (11.7% vs. 11.4%; HR 1.03; 95% CI, 0.95−1.11). This study did not provide evidence that either treatment was superior to the other for secondary stroke prevention.

Cilostazol

Cilostazol inhibits platelet aggregation by blocking phosphodiesterase 3, which leads to increased concentration of cAMP in platelets and blood vessels. Cilostazol is indicated for treatment of claudication in patients with PAD, improving absolute maximal walking distance and quality-of-life measures.[15,16] It should not be used in patients with heart failure because other medications in this class have been shown to increase mortality.

Cilostazol has also been studied as an adjunctive therapy for patients with coronary artery disease who have undergone PCI. The HOST-ASSURE (Harmonizing Optimal Strategy for Treatment of Coronary Artery Stenosis − Safety and Effectiveness of Drug-Eluting Stents and Anti-Platelet Regimen) study randomized 3755 patients undergoing PCI to 1 month of either double-dose clopidogrel dual antiplatelet therapy (DDAT) or triple antiplatelet therapy (TAT) with the addition of cilostazol.[17] In this noninferiority study, the primary endpoint, i.e., a composite of cardiac death, nonfatal MI, stent thrombosis, stroke, and major bleeding, was similar in the two groups (1.4% in DDAT vs. 1.2% in TAT; $p = .0007$ for noninferiority; HR 0.85, 95% CI, 0.49−1.48; $p = .558$ for superiority). Although this study suggested noninferiority, its small size and the availability of newer, more potent antiplatelet agents for patients who are at high risk of stent thrombosis after PCI have prevented the widespread use of cilostazol for coronary artery disease.[18]

Ticagrelor

Ticagrelor is a cyclopentyl-triazolo-pyrimidine (CPTP), a new chemical class of antiplatelet agents that differs from all the thienopyridines described above and causes reversible platelet inhibition, but it is similar to the abovementioned agents in that it inhibits platelet aggregation through the P2Y12 ADP receptor. Unlike the thienopyridines, ticagrelor does not require metabolic activation and it reversibly inhibits the ADP receptor. Because it does not require metabolic activation, it provides faster and more potent platelet inhibition.[19]

The PLATO (Platelet Inhibition and Patient Outcomes) study randomized 18,624 patients admitted to the hospital with ACSs (with or without ST-segment elevation) to either ticagrelor (180 mg loading dose followed by 90 mg twice daily thereafter) or clopidogrel (300–600 mg loading dose followed by 75 mg daily thereafter) added to the background of aspirin therapy and examined the primary endpoint, i.e., a composite of death from vascular causes, MI, or stroke, at 12 months.[20] Compared with patients receiving clopidogrel, those receiving ticagrelor had a lower rate of the primary endpoint (9.8% vs. 11.7%, HR 0.64; 95% CI, 0.77–0.92; $p < .001$), predominantly due to a lower rate of MI (5.8% vs. 6.9%, $p = .005$) and death from vascular causes (4.0% vs. 5.1%, $p = .001$), but not stroke (1.5% vs. 1.3%, $p = .22$). While the rate of TIMI (Thrombolysis in Myocardial Infarction) non-CABG-related major bleeding (including fatal intracranial bleeding) was higher in those receiving ticagrelor (4.5% vs. 3.8%, $p = .03$), all-cause mortality was reduced with ticagrelor (4.5% vs. 5.9%, $p < .001$). This trial, similar to many of the ones discussed in this chapter, showed that an improvement in ischemic risk was achieved at the expense of increased bleeding risk, while all-cause mortality was reduced.

The optimal balance between reduction in ischemic risk and increase in bleeding risk over time was further investigated in the PEGASUS-TIMI 54 (Prevention of Cardiovascular Events in Patients with Prior Heart Attack Using Ticagrelor Compared to Placebo on a Background of Aspirin-Thrombolysis in Myocardial Infarction) study. In this study, 21,162 patients who had an MI 1–3 years earlier were randomized to ticagrelor 90 mg twice daily, 60 mg twice daily, or placebo, when added to a background of low-dose aspirin.[21] Patients were followed up for a median of 33 months. The primary efficacy endpoint, i.e., a composite of cardiovascular death, MI, or stroke, was found to be lower at 3 years in the group receiving ticagrelor compared with the groups receiving placebo and was numerically lowest in the group receiving low-dose ticagrelor. The Kaplan-Meier rates at 3 years were 7.85% for those receiving 90 mg ticagrelor twice daily, 7.77% for those receiving 60 mg twice daily, and 9.04% for those receiving placebo (90 mg ticagrelor vs. placebo HR 0.85; 95% CI, 0.75–0.96;

$p = .008$; 60 mg ticagrelor vs. placebo HR 0.84; 95% CI, 0.74–0.95; $p = .004$). Rates of major bleeding were higher with ticagrelor than placebo (2.6% with 90 mg, 2.3% with 60 mg, 1.06% with placebo, $p < .001$ for each dose vs. placebo), but rates of intracranial hemorrhage or fatal bleeding were similar in the three groups (0.63%, 0.71%, and 0.60%, respectively). Compared with patients receiving ticagrelor 90 mg twice daily, those receiving 60 mg twice daily had lower rates of drug discontinuation (90 mg twice daily, 32.0%; 60 mg twice daily, 28.7%; placebo, 21.4%; $p > .001$ for each ticagrelor dose vs. placebo) and lower rates of dyspnea (90 mg twice daily, 18.9%; 60 mg twice daily, 15.8%; placebo, 6.4%; $p > .001$ for each ticagrelor dose vs. placebo). This trial showed that prolonged duration of DAPT could further improve ischemic risk and that a lower dose over a longer period could be better.

While longer duration DAPT had been studied, only recently did very short duration DAPT undergo investigation in clinical trials powered to assess efficacy. Furthermore, previous studies assessed the benefit of a thienopyridine when added to aspirin versus aspirin alone and not the benefit of aspirin when added to a thienopyridine versus a thienopyridine alone. In the TWILIGHT (Ticagrelor with Aspirin or Alone in High-risk Patients after Coronary Intervention) study, investigators randomized 7119 patients who were at high risk for bleeding or an ischemic event 3 months after PCI who did not suffer a major bleeding or ischemic event to continued ticagrelor plus either aspirin or placebo for 1 year.[22] This study excluded patients with an ST-elevation MI. The primary endpoint of major bleeding was observed in fewer patients receiving ticagrelor plus placebo than in patients receiving ticagrelor plus aspirin (4.0% vs. 7.1%, HR 0.56, 95% CI, 0.45–0.68, $p < .001$). A prespecified composite endpoint of death, nonfatal MI, or nonfatal stroke was 3.9% in both groups. Of note, this trial was not powered to detect differences in stent thrombosis or stroke, which are both clinically important endpoints. Additionally, results of this study are only applicable to patients who were able to tolerate 3 months of DAPT without any major side effects. This study showed that among high-risk patients undergoing PCI, after 3 months of DAPT, it is safe to continue with ticagrelor monotherapy for 1 year and there is less bleeding.

Conclusions

DAPT has become an integral component of the care of patients with vascular disease. Over the past 30 years, investigators have found that although single antiplatelet therapy was beneficial to patients with ACS, DAPT provided further reduction in ischemic risk. Dual pathway platelet inhibition has been studied with a number of agents to achieve an optimal profile that balances ischemic risk with bleeding risk. This involves the potency of the antiplatelet agent, duration of antiplatelet therapy, and patient and plaque characteristics.

References

1. Collaboration AT. Secondary prevention of vascular disease by prolonged antiplatelet treatment. Antiplatelet Trialists' Collaboration. *Br Med J*. 1988;296(6618):320−331. https://www.ncbi.nlm.nih.gov/pubmed/3125883.
2. Randomised trial of intravenous streptokinase, oral aspirin, both, or neither among 17 187 cases of suspected acute myocardial infarction: ISIS-2. *Lancet*. 1988;332(8607): 349−360. https://doi.org/10.1016/S0140-6736(88)92833-4.
3. Mann DL, Libby P, Bonow RO, Eugene Braunwald DPZ, eds. *Braunwald's Heart Disease: A Textbook of Cardiovascular Medicine*. 10th ed. Philadelphia, PA: Elsevier/Saunders; 2015. https://search.library.wisc.edu/catalog/9910209018402121.
4. Leon MB, Baim DS, Popma JJ, et al. A clinical trial comparing three antithrombotic-drug regimens after coronary-artery stenting. *N Engl J Med*. 1998;339(23):1665−1671. https://doi.org/10.1056/NEJM199812033392303.
5. Love BB, Biller J, Gent M. Adverse haematological effects of ticlopidine. *Drug Saf*. 1998;19(2):89−98. https://doi.org/10.2165/00002018-199819020-00002.
6. Page Y, Zeni F, Tardy B, Comtet C, Bertrand JC, Terrena R. Thrombotic thrombocytopenic purpura related to ticlopidine. *Lancet*. 1991;337(8744):774−776. https://doi.org/10.1016/0140-6736(91)91383-6.
7. A randomised, blinded, trial of clopidogrel versus aspirin in patients at risk of ischaemic events (CAPRIE). *Lancet*. 1996;348(9038):1329−1339. https://doi.org/10.1016/S0140-6736(96)09457-3.
8. Effects of clopidogrel in addition to aspirin in patients with acute coronary syndromes without ST-segment elevation. *N Engl J Med*. 2001;345(7):494−502. https://doi.org/10.1056/NEJMoa010746.
9. Mega JL, Close SL, Wiviott SD, et al. Cytochrome P-450 polymorphisms and response to clopidogrel. *N Engl J Med*. 2009;360(4):354−362. https://doi.org/10.1056/NEJMoa0809171.
10. Wallentin L, Varenhorst C, James S, et al. Prasugrel achieves greater and faster P2Y 12 receptor-mediated platelet inhibition than clopidogrel due to more efficient generation of its active metabolite in aspirin-treated patients with coronary artery disease. *Eur Heart J*. 2007;29(1):21−30. https://doi.org/10.1093/eurheartj/ehm545.
11. WS D, AE M, WK J, et al. Randomized comparison of prasugrel (CS-747, LY640315), a novel thienopyridine P2Y12 antagonist, with clopidogrel in percutaneous coronary intervention. *Circulation*. 2005;111(25):3366−3373. https://doi.org/10.1161/CIRCULATIONAHA.104.502815.
12. Wiviott SD, Braunwald E, McCabe CH, et al. Prasugrel versus clopidogrel in patients with acute coronary syndromes. *N Engl J Med*. 2007;357(20):2001−2015. https://doi.org/10.1056/NEJMoa0706482.
13. Aspirin plus dipyridamole versus aspirin alone after cerebral ischaemia of arterial origin (ESPRIT): randomised controlled trial. *Lancet*. 2006;367(9523):1665−1673. https://doi.org/10.1016/S0140-6736(06)68734-5.
14. Sacco RL, Diener H-C, Yusuf S, et al. Aspirin and extended-release dipyridamole versus clopidogrel for recurrent stroke. *N Engl J Med*. 2008;359(12):1238−1251. https://doi.org/10.1056/NEJMoa0805002.
15. Robless P, Mikhailidis DP, Stansby GP. Cilostazol for peripheral arterial disease. *Cochrane Database Syst Rev*. 2008. https://doi.org/10.1002/14651858.CD003748.pub3.

16. Hiatt WR, Money SR, Brass EP. Long-term safety of cilostazol in patients with peripheral artery disease: the CASTLE study (Cilostazol: a study in long-term effects). *J Vasc Surg*. 2008. https://doi.org/10.1016/j.jvs.2007.10.009.

17. Kim HS, Park KW, Kang SH, et al. Adjunctive cilostazol versus double-dose clopidogrel after drug-eluting stent implantation: the HOST-ASSURE randomized trial (harmonizing optimal strategy for treatment of coronary artery stenosis-safety & effectiveness of drug-eluting stents & anti-platelet regimen). *JACC Cardiovasc Interv*. 2013. https://doi.org/10.1016/j.jcin.2013.04.022.

18. Lavie CJ, Dinicolantonio JJ. Cilostazol — a forgotten antiplatelet agent, but does it even matter? *JACC Cardiovasc Interv*. 2013. https://doi.org/10.1016/j.jcin.2013.06.006.

19. James S, Åkerblom A, Cannon CP, et al. Comparison of ticagrelor, the first reversible oral P2Y12 receptor antagonist, with clopidogrel in patients with acute coronary syndromes: rationale, design, and baseline characteristics of the PLATelet inhibition and patient Outcomes (PLATO) trial. *Am Heart J*. 2009. https://doi.org/10.1016/j.ahj.2009.01.003.

20. Wallentin L, Becker RC, Budaj A, et al. Ticagrelor versus clopidogrel in patients with acute coronary syndromes. *N Engl J Med*. 2009;361(11):1045−1057. https://doi.org/10.1056/NEJMoa0904327.

21. Bonaca MP, Bhatt DL, Cohen M, et al. Long-term use of ticagrelor in patients with prior myocardial infarction. *N Engl J Med*. 2015;372(19):1791−1800. https://doi.org/10.1056/NEJMoa1500857.

22. Mehran R, Baber U, Sharma SK, et al. Ticagrelor with or without aspirin in high-risk patients after PCI. *N Engl J Med*. 2019;381(21):2032−2042. https://doi.org/10.1056/NEJMoa1908419.

Acute coronary syndromes and percutaneous coronary intervention—current recommendations for dual antiplatelet therapy. Are guidelines reflecting the data?

Lauren S. Ranard, MD [1], Sorin J. Brener, MD [2]

[1]*Columbia University Medical Center-NewYork Presbyterian Hospital, New York, NY, United States;* [2]*Professor of Medicine, NewYork Presbyterian-Brooklyn Methodist Hospital, Brooklyn, NY, United States*

Introduction

Platelet activation plays a critical role in atherosclerosis and acute coronary syndrome (ACS), and hence antiplatelet therapy is critical to treatment. The foundation of antiplatelet strategy for ACS is aspirin in combination with $P2Y_{12}$ receptor inhibitor therapy. The mechanism of action of these agents has been discussed in detail in earlier chapters.

Advancement in stent technology has been a key achievement in improving clinical outcomes. Since the bare metal stent was introduced in the 1980s, the technology has evolved—first with introduction of a durable polymer to which an eluting antineoplastic drug was affixed (drug-eluting stents [DESs]) and subsequently with the replacement of the durable polymer by one that is bioresorbable. Some DESs even have the antirestenotic agent loaded on the metal without a polymer at all. The implementation of thinner stent struts, thinner polymer coats, and newer antiproliferative/antineoplastic drugs has reduced rates of restenosis and thrombotic complications.[1,2] These advancements have led to faster re-endothelization, allowing for shorter dual antiplatelet therapy (DAPT) duration to be considered.

The US guidelines by the American College of Cardiology (ACC)/American Heart Association (AHA) recommend 12-month DAPT after an ACS both in patients managed medically and in those who received percutaneous coronary intervention

(PCI), and the European guidelines by the European Society of Cardiology (ESC) recommend 6- to 12-month DAPT. There are several randomized trials and meta-analyses that have examined the pharmacotherapy strategy, creating a large amount of data on the topic. Chapters 7 and 8 will address DAPT duration at length. Herein, we will focus on the current evidence for (1) P2Y$_{12}$ inhibitor selection, (2) de-escalation of therapy, and (3) platelet function and genotype testing. Thereafter, we will discuss the current clinical practice guidelines and compare them to the evidence-base that has been generated by decades of studies.

Rationale for the standard 12-month dual antiplatelet therapy

The concept that DAPT is necessary to prevent stent thrombosis and recurrent clinical events has been dominant since the 2000s. The standard 12-month DAPT for ACS is primarily based on results from large, event-driven randomized clinical trials. The foundation was laid by the Clopidogrel in Unstable Angina to Prevent Recurrent Events (CURE) study that randomized patients with ACS without ST-segment elevation to aspirin in addition to either clopidogrel or placebo and demonstrated that the addition of clopidogrel reduces 12-month clinical events such as cardiovascular (CV) death, myocardial infarction (MI), or stroke by ~20%.[3,4] Initial efforts to reduce DAPT duration have been prevented due to the fear of late (>30 days) and very late (>1 year) stent thrombosis.[5]

Randomized trials assessing antiplatelet therapy
Randomized trials assessing P2Y$_{12}$ inhibitor choice

Concerns about interindividual variations in response to clopidogrel led to the development of more potent and rapidly acting agents such as prasugrel and ticagrelor. Overall, these newer agents have the benefit of reducing ischemic events, at the expense of increased bleeding.

The Trial to Assess Improvement in Therapeutic Outcomes by Optimizing Platelet Inhibition with Prasugrel-Thrombolysis in Myocardial Infarction 38 (TRITON-TIMI 38) demonstrated a net clinical benefit of DAPT with prasugrel over clopidogrel in patients with moderate- to high-risk ACS. A 19% reduction in CV death, nonfatal MI, or nonfatal stroke (9.9% vs. 12.1%, hazard ratio [HR] 0.81, 95% confidence interval [CI], 0.73–0.90, $P < .001$) was seen at a median 14.5-month follow-up. This did come at the cost of an increase in life-threatening (1.4% vs. 0.9%, $P = .01$) and fatal bleeding events (0.4% vs. 0.1%, $P = .002$) in those taking prasugrel therapy. The benefit of prasugrel therapy in reducing nonfatal MI was more pronounced in patients with ST-elevation MI (STEMI) than in those with non-ST-elevation MI (NSTEMI) or unstable angina (21% vs. 18% risk reduction), but without the statistical heterogeneity of effect. Moreover, the benefit of prasugrel was accentuated in patients with diabetes (30% risk reduction), with creatinine clearance ≥60 mL/min (14% risk reduction), and aged <65 years (25% risk reduction).[6] The reduction in events occurred early after randomization and persisted through follow-up.

The PLATelet Inhibition and patient Outcomes (PLATO) trial demonstrated the superiority of ticagrelor compared to clopidogrel therapy in patients with NSTEMI, STEMI, and high-risk unstable angina for the prevention of CV ischemic events or death, without causing a significant increase in bleeding. A 16% risk reduction in the combined endpoint of CV death, MI, or stroke was seen at 12 months with the use of ticagrelor compared to clopidogrel therapy (9.8% vs. 11.7%, $HR = 0.84$ [0.77−0.92], $P < .0001$). Importantly, ticagrelor reduced CV mortality by 21% at 1-year (4.0% vs. 5.1%, $HR = 0.79$ [0.69−0.91], $P = .001$). Overall, the trial demonstrated that for every 1000 patients admitted with ACS, the use of ticagrelor therapy for 12 months will result in 14 fewer deaths, 11 fewer MI events, and 6−9 fewer cases of stent thrombosis.[7] Interestingly, ticagrelor therapy was not associated with a significant risk reduction in patients with unstable angina.

The Prasugrel Versus Ticagrelor in Patients with Acute Myocardial Infarction Treated with Primary Percutaneous Coronary Intervention (PRAGUE-18) trial showed similar safety and efficacy between prasugrel and ticagrelor in patients with STEMI. The trial was halted prematurely for futility after interim analysis showed no difference in the primary endpoint between the two antiplatelet agents. The primary endpoint, a composite of both efficacy and safety, included death, urgent target vessel revascularization, stroke, and serious bleeding requiring or prolonging hospitalization at 7 days.[8] At 30 days, there was again no difference in bleeding events between patients according to Thrombolysis in Myocardial Infarction (TIMI) and Bleeding Academic Research Consortium (BARC) criteria. The 12-month intention-to-treat analysis also found no difference in CV death, MI, or stroke (6.6% vs. 5.7%, $P = .503$) or in bleeding events (10.9% vs. 11.1%, $P = .930$). Notably, a large percentage of patients discontinued the study drug and switched to clopidogrel (34.1% of patients who were randomized to prasugrel and 44.4% of patients randomized to ticagrelor). This was more likely due to drug coverage constraints than due to medical considerations. The median time to switching was 8 days, thereby reducing the ability of the study to detect any long-term difference between the two agents.[9] The Intracoronary Stenting and Antithrombotic Regimen: Rapid Early Action for Coronary Treatment 5 (ISAR-REACT 5) trial also compared ticagrelor and prasugrel; however, the results provide an opposing conclusion. ISAR-REACT 5 randomized all patients with ACS and the delivery strategy depended on patient presentation. For patients with STEMI, study drugs were initiated at the time of randomization, whereas in patients with NSTEMI or unstable angina, only ticagrelor was given at randomization and prasugrel was given at the time of angiography and PCI, in accord with its recommendation for use. Prasugrel had superior efficacy to ticagrelor at 1 year with respect to the composite of death, MI, or stroke (6.9% vs. 9.3%, $P = .006$) without affecting the rate of bleeding.

Notably, there was no significant difference in definite or probable stent thrombosis (1.0% vs. 1.3%, HR 1.30, 95% CI, 0.72–2.33). There are some notable criticisms of this landmark trial, including high discontinuation rate at 1 year (15.2% for ticagrelor vs. 12.5% for prasugrel, $P = .03$), open-label nature of the trial, and difference in drug initiation strategy between the agents.[10]

The Ticagrelor in Patients with ST-Elevation Myocardial Infarction Treated with Pharmacological Thrombolysis (TREAT) trial showed that administration of ticagrelor after fibrinolysis for STEMI was noninferior to clopidogrel for major bleeding in patients aged <75 years. Patients were randomized to a median of 11 hours after fibrinolysis and 90% had been pretreated with clopidogrel prior to randomization. The trial additionally demonstrated no difference in CV mortality, MI, stroke/transient ischemic attack, recurrent ischemia, or other arterial thrombotic events at 12 months (8.0% vs. 9.1%, $P = .25$); however, it was not powered for efficacy.[11]

A recently published large network meta-analysis by Navarese et al. compares clopidogrel, ticagrelor, and prasugrel therapy in 52,816 patients with ACS from 12 randomized controlled trials (RCTs) and examined mortality, bleeding, and ischemic outcomes. Compared to clopidogrel, only ticagrelor significantly reduced CV mortality (HR 0.82; CI, 0.72–0.92) and all-cause mortality (HR 0.83; 95% CI, 0.75–0.92). Prasugrel did however significantly reduce recurrent MI compared to clopidogrel (HR 0.81; 95% CI, 0.67–0.98). Both prasugrel and ticagrelor reduced stent thrombosis risk compared to clopidogrel (prasugrel: HR 0.50, 95% CI, 0.38–0.64; ticagrelor: HR 0.72, 95% CI, 0.58–0.90), but both also significantly increased bleeding (prasugrel: HR 1.26, 95% CI, 1.01–1.56; ticagrelor: HR 1.27, 95% CI, 1.04–1.55).[12] Table 4.1 provides a summary of the design and results of the above-described RCTs, while Fig. 4.1 provides a summary of pooled HRs of clinical outcomes from the 12 included trials.

Overall, these key trials suggest that prasugrel is superior to clopidogrel in the prevention of ischemic events, especially in young patients (<65 years old) with STEMI, at the expense of increased bleeding. Ticagrelor, however, is also superior to clopidogrel in the prevention of ischemic events, notably death, with some data suggesting fewer bleeding events. Lastly, the data we have regarding the comparison of ticagrelor and prasugrel $P2Y_{12}$ inhibition are inconclusive and require further investigation.

Randomized trials assessing aspirin dosing or aspirin-free strategies

Sole inhibition of the $P2Y_{12}$ pathway without aspirin is discussed at length in Chapter 10. Both the GLOBAL LEADERS (A Clinical Study Comparing Two Forms of Anti-platelet Therapy after Stent Implantation) and TWILIGHT (Ticagrelor With Aspirin or Alone in High-Risk Patients after Coronary Intervention) trials have investigated aspirin-free strategies for patients with ACS (a subset of the trials) and demonstrated no difference in ischemic events, but a benefit in terms of safety. The GLOBAL LEADERS trial enrolled 15,991 patients with stable coronary artery disease (CAD) or ACS who received PCI and randomized them to either 1 month of

Table 4.1 Randomized trials assessing P2Y$_{12}$ inhibitor choice.

Study name (year)	Patient population	Antiplatelet treatment		N	Outcomes		
		Treatment strategy 1	Treatment strategy 2		Primary efficacy endpoint and result	Primary safety endpoint and result	
TRITON-TIMI 38 (2007)[6]	ACS	Prasugrel	Clopidogrel	13,608	Composite CV death, MI, or stroke at 15 mo.	9.9% versus 12.1%, $P < .001$	Non-CABG-related TIMI major bleeding at 15 mo.	2.4% versus 1.8%, $P = .03$
PLATO (2009)[7]	ACS	Ticagrelor	Clopidogrel	18,624	Composite CV death, MI, or stroke at 12 mo.	9.8% versus 11.7%, $P < .001$	TIMI major bleeding at 12 mo.	11.6% versus 11.2%, $P = .43$
PRAGUE-18 (2016)[8]	STEMI or high-risk NSTEMI	Prasugrel[a]	Ticagrelor	1230	Composite death, reinfarction, TVR, CVA, bleeding requiring transfusion/prolonged hospitalization at 7 d.	4.0% versus 4.1%, $P = .94$	N/A	N/A
TREAT (2018)[11]	Post-fibrinolysis STEMI in patients aged <75 years	Ticagrelor	Clopidogrel	3799	Composite CV death, MI, or CVA at 30 d.	6.7% versus 7.3%, $P = .53$	TIMI major bleeding at 12 m.	1.0% versus 1.2%, $P = .61$
ISAR REACT 5 (2019)[10]	ACS	Ticagrelor	Prasugrel[a]	4018	Composite death, MI, or CVA at 1 y.	9.3% versus 6.9%, $P = .006$	BARC 3–5 at 1 y.	5.4% versus 4.8%, $P = .46$

ACS, acute coronary syndrome; BARC, Bleeding Academic Research Consortium; CABG, coronary artery bypass graft; CV, cardiovascular; CVA, cerebrovascular accident; MI, myocardial infarction; NSTEMI, non-ST-elevation myocardial infarction; STEMI, ST-elevation myocardial infarction; TIMI, Thrombolysis in Myocardial Infarction; TVR, target vessel revascularization.

[a] In patients aged >75 years or weighing <60 kg, the maintenance dose of prasugrel was reduced to 5 mg once daily.

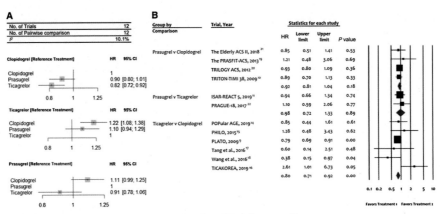

FIGURE 4.1 Summary of pooled hazard ratios (HRs) for mortality.

CI, confidence interval.

Reproduced with permission from Navarese E.P., et al., Comparative efficacy and safety of oral P2Y12 inhibitors in acute coronary syndrome: network meta-analysis of 52,816 patients from 12 randomized trials. Circulation, 2020;142 (2):155–160.

DAPT followed by 23 months of ticagrelor or standard 12-month DAPT followed by aspirin alone.[13] The post hoc analysis restricted to ACS patients, which included 7487 patients, found that between 31 and 365 days after randomization the primary outcome of all-cause death or new Q-wave MI was similar between the two groups (1.5% vs. 2.0%, respectively, HR = 0.73, $P = .07$).[13] More notably, the primary safety endpoint of BARC type 3 or 5 bleeding occurred in significantly fewer patients in the experimental group (0.8% vs. 1.5%, HR 0.52, $P = .004$).[13] Similar results were seen at 2-year follow-up.[14] The TWILIGHT trial enrolled only patients undergoing PCI who were at high risk for ischemic or hemorrhagic complications and had already completed 3 months of DAPT with aspirin and ticagrelor.[15] Thereafter, they were randomized to either ticagrelor monotherapy or DAPT until 12-month follow-up. Out of the 7119 patients enrolled in this trial, 64.8% (4614) were patients with ACS. The subgroup analysis restricted to patients with ACS again demonstrated no difference in all-cause death, nonfatal MI, or stroke between the groups (HR 0.97; 95% CI, 0.73–1.28; $P = .84$); however, it did show a significant reduction in the primary endpoint of BARC 2, 3, or 5 bleeding in those who received ticagrelor monotherapy (3.6% vs. 7.6%, $P < .01$).[15]

The SMART-CHOICE (Comparison Between P2Y12 Antagonist Monotherapy and Dual Antiplatelet Therapy after DES) study also investigated short-duration (90 days) DAPT followed by $P2Y_{12}$ monotherapy compared to standard 12-month therapy and demonstrated noninferiority of the former for major adverse cardiac, cerebrovascular, and bleeding events in the ACS subgroup (3.0% vs. 2.9%, HR 1.06, 95% CI, 0.61–1.85) similar to the non-ACS cohort ($p_{INT} = 0.52$). This study enrolled 2993 patients from 33 Korean centers, of which 58.2% (1741) presented with ACS. Clopidogrel was the predominant agent used.[16] Bleeding (BARC classes 2–5) was substantially reduced in the monotherapy arm.

The TICO (Ticagrelor Monotherapy After 3 Months in the Patients Treated With New Generation Sirolimus Stent for Acute Coronary Syndrome) study is the only trial thus far to look at aspirin-free strategies solely in an ACS population of over 3000 patients. This trial was designed like TWILIGHT and randomized patients at 3 months to either ticagrelor monotherapy or an additional 9 months of DAPT. Ticagrelor monotherapy after 3-month\DAPT was superior to 12-month DAPT in the prevention of net adverse clinical events (death, MI, stent thrombosis, cerebrovascular accident, target vessel revascularization, TIMI major bleeding) at 1 year (3.9% vs. 5.9%, $P = .01$).[17] Landmark analysis at 90 days clearly demonstrated the profound reduction in events after randomization driven primarily by lower bleeding rates. Interestingly, even the ischemic event rates favored monotherapy after 3 months (1.2% vs. 2.1%, HR 0.58, 95% CI, 0.33−1.04; $P = .07$). The ongoing MASTER DAPT (Management of High Bleeding Risk Patients Post Bioresorbable Polymer Coated Stent Implantation With an Abbreviated vs. Prolonged DAPT Regimen) trial will further clarify the net benefit of an aspirin-free strategy as maintenance therapy (NCT03023020).[18] The estimated study completion date is March 2021. Table 4.2 summarizes the design of the trials and results specific to the ACS population of included patients.

The optimal aspirin dose is also an area of interest. Two ongoing trials are evaluating this: ADAPTABLE (Aspirin Dosing: A Patient-centric Trial Assessment Benefits and Long-term Effectiveness) and ANDAMAN (Aspirin Twice a Day in Patients with Diabetes and Acute Coronary Syndrome). ADAPTABLE (NCT02697916) is examining low-dose (81 mg) versus high-dose (325 mg) aspirin in patients with established CAD, with final follow-up estimated to be completed in June 2020.[19] ANDAMAN (NCT02520921) specifically tests aspirin daily versus twice-daily strategies in diabetic patients and results are expected to be available in December 2021.

Randomized trials of de-escalation, platelet function, and genotype types

Similar to trials that have examined the de-escalation of DAPT to aspirin-free strategies in patients with ACS, the Timing of Platelet Inhibition After Acute Coronary Syndrome (TOPIC) trial examined de-escalation to less potent $P2Y_{12}$ inhibition. The TOPIC trial randomized patients with ACS who were event-free after 1 month of initial treatment with DAPT with newer $P2Y_{12}$ antagonists (i.e., ticagrelor or prasugrel) to either de-escalation to DAPT with clopidogrel or continuation on the initial regimen. De-escalation of DAPT was associated with a reduction in bleeding (4.0% vs. 14.9%, $P < .01$) without any difference in ischemic events (9.3% vs. 11.3%, $P = .36$).[20]

Platelet function testing (PFT) initially generated great interest, as it helped identify patients at high risk for recurrent ischemic events. A prospective, multicenter registry, the Assessment of Dual AntiPlatelet Therapy with Drug Eluting Stents (ADAPT-DES) correlated platelet reactivity on clopidogrel-based DAPT with clinic

Table 4.2 Randomized trials assessing aspirin dosing or aspirin-free strategies in ACS.

| Study name (year) | Patient population | Antiplatelet treatment | | N | Outcomes | | |
		Strategy 1	Strategy 2		Primary efficacy endpoint and result	Primary safety endpoint and result	
TICO (2020)[17]	PCI for ACS	ASA + ticagrelor (3 mo) → ticagrelor (9 mo)	ASA + ticagrelor (1 y)	3056	Composite death, MI, ST, CVA, TVR at 1 y.	1.2% versus 2.0% ($P = .09$)	TIMI major bleeding at 1 y.	1.7% versus 3.0% ($P = .02$)
SMART-CHOICE (2019)[16]	PCI with DES	ASA + P2Y$_{12}$ (3 mo) → P2Y$_{12}$ (9 mo)	ASA + P2Y$_{12}$ (1 y)	1741	Composite death, MI, CVA at 1 y.	3.0% versus 2.9 (HR 1.06, 95% CI, 0.61 −1.85)	BARC 2 −5 at 1 y.	1.8% versus 3.2% (HR 0.56, 95% CI, 0.30 −1.05)
TWILIGHT (2019)[15]	PCI with DES[a]	Ticagrelor (1 y)	ASA + ticagrelor (1 y)	4614	Composite all-cause mortality, MI, or CVA at 1 y.	4.3% versus 4.5% (HR 0.97, 95% CI, 0.73 −1.28)	BARC 2, 3, or 5 at 1 y.	3.6% versus 7.6% (HR 0.47, 95% CI, 0.36 −0.61)
GLOBAL LEADERS (2018)[14]	PCI for stable or unstable CAD	ASA + ticagrelor (1 mo) → ticagrelor (23 mo)	ASA + ticagrelor/ clopidogrel (1 y) → ASA (1 y)	7487	Composite all-cause mortality, or new nonfatal centrally adjudicated Q-wave MI at 2 y.	3.9% versus 4.5% ($P = .19$)	BARC 3 or 5 at 2 y.	1.9% versus 2.7% ($P = .037$)

The results included in this table are specific to the ACS population in the abovementioned trials.

ACS, acute coronary syndrome; ASA, aspirin; BARC, Bleeding Academic Research Consortium; CAD, coronary artery disease; CI, confidence interval; CVA, cerebrovascular accident; DES, drug-eluting stent; HR, hazard ratio; MI, myocardial infarction; PCI, percutaneous coronary intervention; ST, stent thrombosis; TIMI, Thrombolysis in Myocardial Infarction; TVR, target vessel revascularization.

[a] Eligible patients had completed 3 mo of dual antiplatelet therapy with ticagrelor after index PCI, had ≥1 high-risk feature of ischemia or bleeding, and had not experienced any major bleeding or ischemic event since index PCI.

outcomes.[21] The registry included 8582 patients at both US and European institutions treated with one or more DESs and tested platelet reactivity after DES placement using the VerifyNow POC assay (Accriva Diagnostics). High aspirin platelet reactivity ($\geq$550 aspirin reaction units), with or without high clopidogrel platelet reactivity (>208 $P2Y_{12}$ reaction units), was not associated with clinical outcomes.[22] High clopidogrel reactivity (>208 $P2Y_{12}$ reaction units), however, was an independent predictor of 1-year stent thrombosis and MI (HR 2.49, 95% CI, 1.43–4.31; $P < .001$) after DES placement. This conflicts with findings from the Intracoronary Stenting and Antithrombotic Regimen-Aspirin and Platelet Inhibition (ISAR-ASPI) registry, which demonstrated high aspirin platelet reactivity to be associated with a greater risk of ischemic events at 1 year.[23] Methodologically, the ISAR-ASPI registry enrolled only German patients and blood samples were obtained after administration of 500 mg of intravenous aspirin. The Multiplate analyzer was used to test platelet inhibition, with high aspirin platelet reactivity defined as >203 aspirin aggregation units/min. Because of the differences in platelet testing methods and the population studied, the studies are therefore potentially not directly comparable. Notably, a study comparing PFT technology found that only testing utilizing VerifyNow and Plateletworks technology correlated with CV endpoints.[24]

Randomized trials have also examined the utility of PFT to guide antiplatelet therapy. The Is There A Life for DES After Discontinuation of Clopidogrel (ITALIC) trial examined shorter duration DAPT in aspirin-sensitive, low-risk patients (excluded patients with left main PCI, ACS, or prior DES in the past 1 year) who received a DES and demonstrated that 6-month DAPT is noninferior to 24-month DAPT (absolute risk difference 0.11%; for noninferiority, $P = .0002$). Aspirin sensitivity was assessed after the first dose of aspirin; one of the three aspirin resistance tests (PFA-100, Multiplate electrical impedance aggregometry, or VerifyNow) were used. Patients on any $P2Y_{12}$ were included in the trial; however, clopidogrel was the $P2Y_{12}$ of choice in 98.7% of patients. Notably, in the shorter duration arm, 8.9% took DAPT for longer durations and 24.2% did not follow the prescribed 6-month duration. As the trial was terminated early due to low recruitment, it is quite underpowered.[25,26] In the Testing Responsiveness to Platelet Inhibition on Chronic Antiplatelet Treatment For Acute Coronary Syndromes (TROPICAL-ACS) trial, patients were treated with prasugrel for 1 week, then clopidogrel for 1 week, and subsequently underwent PFT. High platelet reactivity on clopidogrel has been linked to stent thrombosis and adverse CV events after stenting, and therefore if high platelet reactivity was documented, patients were switched back to prasugrel; otherwise they remained on clopidogrel for the remainder of the 12-month period.[24,27,28] De-escalation of therapy was found to be noninferior to 12 months of prasugrel in patients with ACS. Notably, there was a significant treatment interaction when the analysis was stratified by age ($P = .03$), such that the net clinical benefit in de-escalation was due to reduction in major bleeding only in younger patients.[29] These results are clinically relevant, as PFT can be incorporated into decision-making when the cost for alternative $P2Y_{12}$ antagonists to clopidogrel is an issue.

Genotype testing is not part of guideline-directed care; however, up to 30% of patients have CYP2C19 loss of function (LOF) variants, making them potentially resistant to clopidogrel therapy, which is the most frequently prescribed $P2Y_{12}$ inhibitor.[30] Platelet inhibition of alternatives to clopidogrel, such as ticagrelor or prasugrel, is much less affected by these genetic variants.[31−33] Multiple studies have been performed demonstrating the association of CYP2C19 LOF alleles with clinical outcomes in clopidogrel-treated patients, demonstrating inconsistent results. One meta-analysis including 9685 patients showed a significantly increased risk of major adverse cardiac events (one LOF allele: HR 1.55, 95% CI, 1.11−2.17; 2 LOF alleles: HR 1.76, 95% CI, 1.24−2.50) and stent thrombosis (one LOF allele: HR 2.67, 95% CI, 1.69−4.22; 2 LOF alleles: HR 3.97, 95% CI, 1.75−9.02) in carriers of CYP2C19 LOF alleles and an increasing risk with a greater number of LOF alleles.[34] A subsequent meta-analysis, however, did not show the CYP2C19 genotype to be significantly associated with CV events. The second meta-analysis was not specific for patients undergoing stenting, whereas the first described meta-analysis included only patients who were receiving clopidogrel after PCI.[35] Needless to say, these conclusions are prone to bias, as genotyping was not performed prospectively and the decision to treat was not based on its results.

The Pharmacogenetics of Clopidogrel in Patients with Acute Coronary Syndromes (PHARMCLO) trial utilized point-of-care genotype testing to identify intermediate and poor clopidogrel metabolizers and demonstrated that the use of genotype testing influenced providers' choice of $P2Y_{12}$ inhibitor after PCI, resulting in fewer ischemic and bleeding events. Patients with ACS were randomized to either genotype-based $P2Y_{12}$ inhibitor or usual care. The trial had a total of 888 enrollees, as it was terminated at <25% of the planned enrollment; therefore, although interesting, these results were obtained from a small, underpowered trial.[36] The Patient Outcome After Primary PCI (POPular) Genetics study randomized patients with STEMI undergoing primary PCI to CYP2C19 genotyping or to routine ticagrelor or prasugrel treatment and examined major adverse cardiac events and bleeding at 1 year. In the genotyping group, patients with ≥1 LOF allele(s) received ticagrelor or prasugrel and all noncarriers received clopidogrel.[37] Overall, genotype-guided treatment was noninferior to standard treatment. Although there was a difference between the groups in bleeding outcomes, this was largely driven by the lower incidence of minor bleeding in the genotype-guided group.[38]

The Tailored Antiplatelet Initiation to Lessen Outcomes due to decreased Clopidogrel Response After PCI (TAILOR-PCI, NCT01742117) trial is the largest cardiology trial to date to examine the effectiveness of using genotype testing (TaqMan) for CYP2C19 to guide treatment choice. This is an international trial that has enrolled 5302 patients after undergoing PCI for ACS or stable CAD. Patients are randomized to clopidogrel 75 mg daily or genotype-guided therapy (ticagrelor or clopidogrel). The trial's estimated completion date is March 2021. Preliminary 1-year data demonstrated that in genetic variant carriers, the primary efficacy endpoint was not statistically different (4% for genotype-guided group vs. 5.9% in the conventional group, $P = .056$). A post hoc analysis found ~80% reduction in adverse events in the first 3 months of treatment in patients who received genotype-guided therapy.[39] Although this study did not meet its goal to reduce ischemic events by 50%, there was a clinically relevant reduction. Additionally, the initial trial results did demonstrate that in the early period after PCI, there is potentially a benefit to tailor treatment with genotype-guided therapy.

Table 4.3 summarizes the design and results of the aforementioned RCTs in this section. The role and clinical utility of pharmacogenetics in guiding antiplatelet treatment remains unresolved. The lack of standardization in testing is a key limitation for its broader application. There is potential utility for PFT in risk stratification of patients for ischemic and bleeding complications, but more data in this setting are necessary.

Current clinical practice guidelines

The ACC/AHA and ESC clinical practice guidelines contain not only large areas of overlap but also some notable differences. The last complete guideline update by the ACC/AHA was published in 2016 and in 2017 for the ESC.[40,41] We will review the recommendations by both guidelines in the following clinical settings: (1) PCI in ACS, (2) stable CAD, and (3) post-CABG (coronary artery bypass graft). Fig. 4.2 reviews the recommendations for the duration of therapy in patients with specifically high bleeding risk. We will also review the recommendations for switching of $P2Y_{12}$ inhibitor therapy and platelet function/pharmacogenetic testing.

Percutaneous coronary intervention for acute coronary syndrome

Recommendations on $P2Y_{12}$ inhibitor selection and duration in patients with NSTEMI and in those with STEMI are largely consistent between the US and European guidelines, with both recommending aspirin and either ticagrelor or prasugrel in preference to clopidogrel (class of recommendation [COR] IIa, level of evidence [LOE] B) in the absence of contraindications or thrombolysis. Patients who receive thrombolysis should receive clopidogrel (COR I, LOE A). Furthermore, both guidelines recommend DAPT for at least 12 month regardless of the type of stent implanted (COR I, LOE B) and consideration for prolongation beyond 12 months in those who have tolerated DAPT without complications (COR IIb, LOE A). Those who have a high bleeding risk or develop bleeding may discontinue DAPT at 6 months (AHA/ACC: COR IIb, LOE C; ESC COR IIa, LOE B).

The recommendations for pretreatment differ slightly between the guidelines. The 2016 ACC/AHA guidelines recommend administration of the loading dose prior to procedure for patients with NSTEMI undergoing PCI (COR 1, LOE A) and as early as possible in patients with STEMI (COR I, LOE B). On the other hand, the 2017 ESC guidelines are specific to the $P2Y_{12}$ inhibitor used. In patients

Table 4.3 Randomized trials of de-escalation, platelet function testing, and genotype testing.

Study name (year)	Antiplatelet strategy				Outcomes			
	Patient population	Arm 1	Arm 2	N	Primary endpoint and result	Primary endpoint and result	Primary safety endpoint and result	Primary safety endpoint and result
ITALIC (2015)[25]	PCI with DES and nonresistant to ASA[a]	ASA + clopidogrel/prasugrel/ticagrelor (6 mo) → ASA	ASA + clopidogrel/prasugrel/ticagrelor (2 y) → ASA	1850	Composite death, MI, emergent TVR, CVA, major bleeding at 12 mo.	1.6% versus 1.5% $P = .85$	N/A	N/A
TOPIC (2017)[20]	ACS requiring PCI and received ASA + ticagrelor/prasugrel at discharge[b]	ASA + ticagrelor/prasugrel (1 y)	ASA + clopidogrel (1 y)	645	Composite CV death, emergent TVR, CVA, and BARC ≥2 at 1 y.	26.3% versus 13.4%, $P < .01$	BARC ≥2 at 1 yr.	14.9% versus 4.0%, $P < .01$
TROPICAL ACS (2017)[44]	ACS requiring PCI and treatment with prasugrel	ASA + prasugrel (1 wk) → ASA + clopidogrel (1 wk) → platelet function measurement guided treatment to either prasugrel or clopidogrel (1 y).[c]	ASA + prasugrel (1 y)	2610	Composite CV death, CVA, MI, BARC ≥2 at 1 y.	7% versus 9%; for noninferiority, $P = 0.004$	BARC ≥2 at 1 y.	5% versus 6%, $P = .23$

Study	Type	Intervention	Comparator	n	Primary endpoint	Result	Secondary endpoint	Result
PHARMCLO (2018)[36]	STEMI or NSTEMI	ASA + $P2Y_{12}$ (clopidogrel, ticagrelor, prasugrel) based on genotype results[d]	ASA + $P2Y_{12}$ based on clinical characteristics alone	888	Composite CV death, MI, CVA, BARC 3–5 at 12 mo.	15.9% versus 25.9%, $P < .001$	BARC 3–5 at 12 mo.	4.2% versus 6.8%, $P = .10$
POPular genetics (2019)[38]	STEMI	Early genotyping if ≥1 LOF allele, then ASA + ticagrelor, otherwise ASA + clopidogrel (1 y)	ASA + prasugrel/ ticagrelor (1 y)	2488	Composite death from any cause, MI, definite ST, CVA, or PLATO major bleeding at 1 y.	5.1% versus 5.9%; for noninferiority, $P < 0.001$	PLATO major and minor bleeding at 1 y.	9.8% versus 12.5%, $P = .04$

ACS, acute coronary syndrome; ASA, aspirin; BARC, Bleeding Academic Research Consortium; CABG, coronary artery bypass graft; CV, cardiovascular; CVA, cerebrovascular accident; DES, drug-eluting stent; LOF, loss of function; MI, myocardial infarction; NSTEMI, non–ST-elevation myocardial infarction; PCI, percutaneous coronary intervention; PLATO, Platelet Inhibition and Patient Outcomes; STEMI, ST-elevation myocardial infarction; TVR, target vessel revascularization.

[a] ASA resistance was assessed after an initial dose of 75 mg. Patients responding poorly to the first ASA dose either were considered resistant or underwent a second check after 2 days of 160 mg oral ASA; a third check was made after 2 days of 325 mg oral ASA in case of poor response to 160 mg, and this dose was applied throughout the trial.

[b] Patients were enrolled if they experienced no major adverse event 1 month after ACS.

[c] If platelet function test at 2 weeks showed high reactivity (≥46 U by ADPtest), then patients were switched back to prasugrel. If they had an adequate response to clopidogrel (<46 U by ADPtest), they were continued on clopidogrel for the remaining 11.5 months of the study.

[d] Patients in the genotyping arm underwent testing using the STQ3 system for ABCB1 3435, CYP2C19*2, and CYP2C19*17.

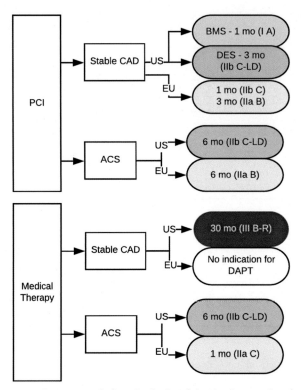

FIGURE 4.2 Summary of recommendations for dual antiplatelet therapy duration in patients at high bleeding risk.

ACS, acute coronary syndrome; *BMS*, bare metal stent; *CAD*, coronary artery disease; *DAPT*, dual antiplatelet therapy; *DES*, drug-eluting stent; *EU*, European Society of Cardiology Guidelines; *PCI*, percutaneous coronary intervention; *US*, United States ACC/AHA guidelines.

with NSTEMI, prasugrel is recommended only for patients whose coronary anatomy is known prior to PCI, otherwise COR III, LOE B is recommended. Conversely, prasugrel can be given before coronary anatomy is known in patients with STEMI.

For patients who receive only medical therapy for ACS, both the 2016 ACC/AHA and 2017 ESC guidelines recommend 12-month DAPT with aspirin and clopidogrel or ticagrelor (ACC/AHA: COR I, LOE B, ESC: COR I, LOE A), with an option to extend beyond 12 months for those at low bleeding risk (ACC/AHA: COR IIb, LOE A; ESC: COR IIb, LOE B (ticagrelor); ESC: COR IIb, LOE C (clopidogrel)). Both guidelines prefer ticagrelor to clopidogrel (ACC/AHA: COR IIa, LOE B; ESC: COR I, LOE B). Prasugrel is not recommended in this setting (COR III, LOE B). The 2017 ESC guidelines also recommend shortening the duration of DAPT to 1 month in those at high bleeding risk (COR IIa, LOE C).

Stable coronary artery disease

For patients with stable CAD undergoing PCI, both US and European guidelines recommend clopidogrel as the drug of choice (ESC: COR I, LOE A; ACC/AHA COR I, LOE B). The 2017 ESC guidelines further state that prasugrel and ticagrelor can be considered in those with high ischemic risk and low bleeding risk (COR IIb, LOE C). These recommendations will be shortly updated, as ticagrelor was recently approved in the United States for non-ACS post-PCI patients based on positive results of the THEMIS (Ticagrelor in Patients with Stable Coronary Disease and Diabetes) trial.[42]

The 2016 AHA/ACC guidelines recommend aspirin indefinitely (COR I, LOE B) and clopidogrel for 1 month after bare metal stent implantation (COR I, LOE A) or 6 months after DES placement (COR I, LOE B).[41] This differs from the 2017 ESC guidelines that recommend 6 months of DAPT irrespective of the stent type (COR I, LOE A) or if patients are treated with a drug-coated balloon angioplasty (COR IIa, LOE B).[40] The 2016 AHA/ACC guidelines further specify that if patients are at high risk for bleeding and received a DES, DAPT can be discontinued at 3 months (COR IIb, LOE C).[41] The 2017 ESC guidelines state the same (COR IIa, LOE B) and even consider shortening the duration of DAPT to 1 month in this situation (COR IIb, LOE C).[40] Additionally, patients who tolerate DAPT are candidates for an undefined period of prolonged DAPT (COR IIb, LOE A).[41]

Coronary artery bypass graft

The 2016 ACC/AHA guidelines state that DAPT must be restarted "as soon as possible" after CABG in patients with ACS or who have had recent stent implantation (COR I, LOE C) and must be maintained for 12 months. DAPT may also be considered in patients with stable CAD who have undergone CABG to improve graft patency (COR IIb, LOE B). Conversely, the 2017 ESC guidelines consider 6-month DAPT as sufficient for those with a prior MI and at high risk for bleeding (COR IIa, LOE C), while those with low bleeding risk may be considered for extension of DAPT to 36 months (COR IIb, LOE C).

Switching of P2Y$_{12}$ inhibitors

The issue of transitioning to P2Y$_{12}$ inhibitors is covered in detail only in the 2017 ESC guidelines. Early intensification of therapy from clopidogrel to ticagrelor therapy is recommended in patients with ACS (COR I, LOE B). A practical algorithm for switching between oral P2Y$_{12}$ inhibitors in the acute or chronic setting is provided in the 2017 ESC guidelines, with administration of a loading dose always recommended in the acute setting.[40] In the chronic setting, a reload is only recommended when transitioning from ticagrelor to prasugrel or from ticagrelor to clopidogrel, due to the differing mechanism of receptor blockade for the two types of inhibitors.[40]

Platelet function testing and genetic testing

The 2018 ESC guidelines for myocardial revascularization incorporate PFT to guide de-escalation of $P2Y_{12}$ inhibitors, especially for those who are high risk for bleeding complications from 12-month DAPT (COR IIb, LOE B).[43] Patients who have a normal clopidogrel platelet inhibition response would be eligible to consider de-escalation from prasugrel to clopidogrel. PFT is incorporated only into the 2017 ESC guidelines to guide timing of CABG in patients who have recently received $P2Y_{12}$ inhibitors (COR IIb, LOE B).

Risk stratification for ischemic and bleeding events

As the average age of the population increases, it is unsurprising that the average number of medical problems in the patients admitted with ACS is increasing, thereby adding to patient complexity. This presents challenges in optimizing DAPT, as both thrombotic and bleeding risk is enhanced. Risk stratification is an overriding concept in both the ACC/AHA and ESC guidelines. There currently exist few tools to aid clinicians with individualizing the decision for duration and de-escalation of DAPT; however, the DAPT score and PRECISE-DAPT are two tools that are available and highlighted in the guidelines. These and other schemes for risk assessment are detailed in Chapter 9.

Risk stratification can be further complicated by the fact that one-third of patients with ACS also have an indication for oral anticoagulation, further increasing hemorrhagic risk and complicating selection of the optimal regimen, as detailed in Chapter 11. Therefore as there are multiple factors that need to be considered, an individualized approach should be taken to determine the duration, combination, and type of $P2Y_{12}$ inhibition in patients who require DAPT.

Do the current guidelines reflect the data?

Antithrombotic pharmacotherapy is a rapidly evolving field, requiring regular updates on recommendations for DAPT. The ACC/AHA and ESC guidelines reflect the currently available data on $P2Y_{12}$ inhibitor selection; however, they have lagged in incorporating when/how to de-escalate treatment and conduct PFT. Table 4.4 compares the major topics in the guidelines to the currently available data.

The results from TWILIGHT, GLOBAL LEADERS, and similar trials were groundbreaking in demonstrating that early de-escalation to $P2Y_{12}$ inhibition monotherapy is noninferior to standard care.[13,15] Importantly, a significant reduction in bleeding with de-escalation of therapy has been demonstrated without affecting ischemic outcomes.[15,20] As risk stratification was an overarching theme in the last set of guidelines, we anticipate that the next updates will include recommendations for de-escalation in the context of bleeding and ischemic risk.

Table 4.4 Comparison of ACC/AHA and ESC guidelines on DAPT to the currently available data.

Topic	2016 ACC/AHA update	2017 ESC update	Current data
Risk stratification	DAPT score to assess for risk/benefit of prolonging therapy	DAPT and PRECISE-DAPT scores to assess risk/ benefit of prolonging therapy	
Type of P2Y$_{12}$ inhibitor in patients with ACS undergoing PCI	- Ticagrelor > clopidogrel (class IIa) - Prasugrel > clopidogrel[a] (class IIa)	- Ticagrelor > clopidogrel (class I) - Prasugrel if P2Y$_{12}$ inhibitor-naïve patients[a,b] (class I)	- Prasugrel > clopidogrel in the prevention of ischemic outcomes at the expense of bleeding (TRITON-TIMI 38) - Ticagrelor > clopidogrel (PLATO)
Type of P2Y$_{12}$ inhibitor in patients with stable CAD undergoing PCI	Clopidogrel (class I)	- Clopidogrel (class I) - Ticagrelor or prasugrel can be considered taking into account ischemia and bleeding risks (class IIb)	
Switching of P2Y$_{12}$ inhibitors	Does not comment; states "no randomized data available on long-term safety or efficacy of switching P2Y12 inhibitors"	- Early upgrade from clopidogrel → ticagrelor in ACS (class I) - Switching between P2Y$_{12}$ inhibitors in case of side effect or intolerance (class IIb)	De-escalation of therapy if no adverse events from ticagrelor/prasugrel → clopidogrel improves bleeding outcomes (TOPIC)
Platelet function testing Genotype testing	Routine use of PFT and genetic testing is not recommended (class III)	Routine PFT to adjust antiplatelet therapy before or after stenting is not recommended (class III, ESC 2017); however, ESC 2018 update allows de-escalation of prasugrel → clopidogrel for those at high risk for bleeding and normal clopidogrel platelet inhibition response.	Inconclusive results Genotype testing helps tailor therapy to improve ischemic outcomes (PHARMCLO)

ACC, *American College of Cardiology*; ACS, *acute coronary syndrome*; AHA, *American Heart Association*; CAD, *coronary artery disease*; DAPT, *dual antiplatelet therapy*; ESC, *European Society of Cardiology*; PCI, *percutaneous coronary intervention*; PFT, *platelet function testing*.

[a] *Prasugrel only recommended if not at high risk for bleeding and no history of stroke or transient ischemic attack.*
[b] *Prasugrel is only recommended in non-ST-elevation myocardial infarction if coronary anatomy is known prior.*

Genotype and PFT to tailor DAPT is not incorporated into the ACC/AHA guidelines and is only selectively incorporated into the 2018 ESC guidelines for myocardial revascularization.[43] There has yet to be robust clinical evidence from randomized trials to guide the use of these modalities. Results from PHARMCLO and preliminary data from TAILOR-PCI suggest that genotype testing influences providers' choice of $P2Y_{12}$ inhibitor, with meaningful clinical consequences.

Conclusion

ACSs affect ~1 million patients yearly. Enhanced platelet activation and aggregation in the setting of plaque destabilization is inexorably linked to the clinical presentation, placing antiplatelet therapy at the center of the therapeutic paradigm. Rigorous clinical trials have resulted in a wealth of data to guide clinical practice, including agent choice and personalization of care based on the assessment of ischemic and bleeding risk. The results of ongoing trials will help further refine this process.

References

1. Piccolo R, Bonaa KH, Efthimiou O, et al. Drug-eluting or bare-metal stents for percutaneous coronary intervention: a systematic review and individual patient data meta-analysis of randomised clinical trials. *Lancet*. 2019;393(10190):2503−2510.
2. Omar WA, Kumbhani DJ. The current literature on bioabsorbable stents: a review. *Curr Atherosclerosis Rep*. 2019;21(12):54.
3. The Clopidogrel in Unstable Angina to Prevent Recurrent Events Trial Investigators. Effects of clopidogrel in addition to aspirin in patients with acute coronary syndromes without ST-segment elevation. *N Engl J Med*. 2001;345(7):494−502.
4. Mehta SR, Yusuf S, Peters RJ, et al. Effects of pretreatment with clopidogrel and aspirin followed by long-term therapy in patients undergoing percutaneous coronary intervention: the PCI-CURE study. *Lancet*. 2001;358(9281):527−533.
5. Spertus JA, Kettelkamp R, Vance C, et al. Prevalence, predictors, and outcomes of premature discontinuation of thienopyridine therapy after drug-eluting stent placement. *Circulation*. 2006;113(24):2803−2809.
6. Wiviott SD, Braunwald E, McCabe CH, et al. Prasugrel versus clopidogrel in patients with acute coronary syndromes. *N Engl J Med*. 2007;357(20):2001−2015.
7. Wallentin L, Becker RC, Budaj A, et al. Ticagrelor versus clopidogrel in patients with acute coronary syndromes. *N Engl J Med*. 2009;361(11):1045−1057.
8. Motovska Z, Hlinomaz O, Miklik R, et al. Prasugrel versus ticagrelor in patients with acute myocardial infarction treated with primary percutaneous coronary intervention: multicenter randomized PRAGUE-18 study. *Circulation*. 2016;134(21):1603−1612.
9. Motovska Z, Hlinomaz O, Kala P, et al. 1-year outcomes of patients undergoing primary angioplasty for myocardial infarction treated with prasugrel versus ticagrelor. *J Am Coll Cardiol*. 2018;71(4):371−381.

10. Schüpke S, Neumann FJ, Menichelli M, et al. Ticagrelor or prasugrel in patients with acute coronary syndromes. *N Engl J Med.* 2019;381(16):1524−1534.

11. Berwanger O, Nicolau JC, Carvalho AC, et al. Ticagrelor vs clopidogrel after fibrinolytic therapy in patients with ST-elevation myocardial infarction: a randomized clinical trial. *JAMA Cardiol.* 2018;3(5):391−399.

12. Navarese EP, Khan SU, Kołodziejczak M, et al. Comparative efficacy and safety of oral P2Y12 inhibitors in acute coronary syndrome: network meta-analysis of 52,816 patients from 12 randomized trials. *Circulation.* 2020;142(2):155−160.

13. Tomaniak M, Chichareon P, Onuma Y, et al. Benefit and risks of aspirin in addition to ticagrelor in acute coronary syndromes: a post hoc analysis of the randomized GLOBAL LEADERS trial. *JAMA Cardiol.* 2019;4(11):1092−1101.

14. Vranckx P, Valgimigli M, Juni P, et al. Ticagrelor plus aspirin for 1 month, followed by ticagrelor monotherapy for 23 months vs aspirin plus clopidogrel or ticagrelor for 12 months, followed by aspirin monotherapy for 12 months after implantation of a drug-eluting stent: a multicentre, open-label, randomised superiority trial. *Lancet.* 2018;392(10151):940−949.

15. Mehran R, Baber U, Sharma SK, et al. Ticagrelor with or without aspirin in high-risk patients after PCI. *N Engl J Med.* 2019;381(21):2032−2042.

16. Hahn J-Y, Song YB, Oh J-H, et al. Effect of P2Y12 inhibitor monotherapy vs dual antiplatelet therapy on cardiovascular events in patients undergoing percutaneous coronary intervention: the SMART-CHOICE randomized clinical trial. *J Am Med Assoc.* 2019;321(24):2428−2437.

17. Kim B-K, Hong S-J, Cho Y-H, et al. Effect of ticagrelor monotherapy vs ticagrelor with aspirin on major bleeding and cardiovascular events in patients with acute coronary syndrome: the TICO randomized clinical trial. *J Am Med Assoc.* 2020;323(23):2407−2416.

18. Frigoli E, Smits P, Vranckx P, et al. Design and rationale of the management of high bleeding risk patients post bioresorbable polymer coated stent implantation with an abbreviated versus standard DAPT regimen (MASTER DAPT) study. *Am Heart J.* 2019;209:97−105.

19. Marquis-Gravel G, Roe MT, Robertson HR, et al. Rationale and design of the aspirin dosing—a patient-centric trial assessing benefits and long-term effectiveness (ADAPT-ABLE) trial. *JAMA Cardiol.* 2020;5(5):598−607.

20. Cuisset T, Deharo P, Quilici J, et al. Benefit of switching dual antiplatelet therapy after acute coronary syndrome: the TOPIC (timing of platelet inhibition after acute coronary syndrome) randomized study. *Eur Heart J.* 2017;38(41):3070−3078.

21. Stone GW, Witzenbichler B, Weisz G, et al. Platelet reactivity and clinical outcomes after coronary artery implantation of drug-eluting stents (ADAPT-DES): a prospective multicentre registry study. *Lancet.* 2013;382(9892):614−623.

22. Chung CJ, Kirtane AJ, Zhang Y, et al. Impact of high on-aspirin platelet reactivity on outcomes following successful percutaneous coronary intervention with drug-eluting stents. *Am Heart J.* 2018;205:77−86.

23. Mayer K, Bernlochner I, Braun S, et al. Aspirin treatment and outcomes after percutaneous coronary intervention: results of the ISAR-ASPI registry. *J Am Coll Cardiol.* 2014;64(9):863−871.

24. Breet NJ, van Werkum JW, Bouman HJ, et al. Comparison of platelet function tests in predicting clinical outcome in patients undergoing coronary stent implantation. *J Am Med Assoc.* 2010;303(8):754−762.

25. Gilard M, Barragan P, Noryani AAL, et al. 6- versus 24-month dual antiplatelet therapy after implantation of drug-eluting stents in patients nonresistant to aspirin: the randomized, multicenter ITALIC trial. *J Am Coll Cardiol*. 2015;65(8):777−786.

26. Didier R, Morice MC, Barragan P, et al. 6- versus 24-month dual antiplatelet therapy after implantation of drug-eluting stents in patients nonresistant to aspirin: final results of the ITALIC trial (is there a life for DES after discontinuation of clopidogrel). *JACC Cardiovasc Interv*. 2017;10(12):1202−1210.

27. Sibbing D, Braun S, Morath T, et al. Platelet reactivity after clopidogrel treatment assessed with point-of-care analysis and early drug-eluting stent thrombosis. *J Am Coll Cardiol*. 2009;53(10):849−856.

28. Parodi G, Marcucci R, Valenti R, et al. High residual platelet reactivity after clopidogrel loading and long-term cardiovascular events among patients with acute coronary syndromes undergoing PCI. *J Am Med Assoc*. 2011;306(11):1215−1223.

29. Sibbing D, Aradi D, Jacobshagen C, et al. A randomised trial on platelet function-guided de-escalation of antiplatelet treatment in ACS patients undergoing PCI. Rationale and design of the testing responsiveness to platelet inhibition on chronic antiplatelet treatment for acute coronary syndromes (TROPICAL-ACS) trial. *Thromb Haemostasis*. 2017;117(1):188−195.

30. Pereira NL, Rihal CS, So DY, et al. Clopidogrel pharmacogenetics: state-of-the-art review and the TAILOR-PCI study. *Circulation: Cardiovascular Interventions*. 2019; 12(4):e007811.

31. Mega JL, Close SL, Wiviott SD, et al. Cytochrome p-450 polymorphisms and response to clopidogrel. *N Engl J Med*. 2009;360(4):354−362.

32. Wallentin L, James S, Storey RF, et al. Effect of CYP2C19 and ABCB1 single nucleotide polymorphisms on outcomes of treatment with ticagrelor versus clopidogrel for acute coronary syndromes: a genetic substudy of the PLATO trial. *Lancet*. 2010;376(9749): 1320−1328.

33. Tantry US, Bliden KP, Wei C, et al. First analysis of the relation between CYP2C19 genotype and pharmacodynamics in patients treated with ticagrelor versus clopidogrel: the ONSET/OFFSET and RESPOND genotype studies. *Circulation: Cardiovascular Genetics*. 2010;3(6):556−566.

34. Mega JL, Simon T, Collet J-P, et al. Reduced-function CYP2C19 genotype and risk of adverse clinical outcomes among patients treated with clopidogrel predominantly for PCI: a meta-analysis. *J Am Med Assoc*. 2010;304(16):1821−1830.

35. Holmes MV, Perel P, Shah T, Hingorani AD, Casas JP. CYP2C19 genotype, clopidogrel metabolism, platelet function, and cardiovascular events: a systematic review and meta-analysis. *J Am Med Assoc*. 2011;306(24):2704−2714.

36. Notarangelo FM, Maglietta G, Bevilacqua P, et al. Pharmacogenomic approach to selecting antiplatelet therapy in patients with acute coronary syndromes: the PHARMCLO trial. *J Am Coll Cardiol*. 2018;71(17):1869−1877.

37. Bergmeijer TO, Janssen PW, Schipper JC, et al. CYP2C19 genotype−guided antiplatelet therapy in ST-segment elevation myocardial infarction patients−rationale and design of the Patient Outcome after primary PCI (POPular) Genetics study. *Am Heart J*. 2014; 168(1), 16−22. e1.

38. Claassens DMF, Vos GJA, Bergmeijer TO, et al. A genotype-guided strategy for oral P2Y12 inhibitors in primary PCI. *N Engl J Med*. 2019;381(17):1621−1631.

39. Pereira NL, Farkouh ME, So D, et al. Effect of Genotype-Guided Oral P2Y12 Inhibitor Selection vs Conventional Clopidogrel Therapy on Ischemic Outcomes After Percutaneous Coronary Intervention. *J Am Med Assoc*. 2020;324(8):761−771.

40. Valgimigli M, Bueno H, Byrne RA, et al. 2017 ESC focused update on dual antiplatelet therapy in coronary artery disease developed in collaboration with EACTS: the Task Force for dual antiplatelet therapy in coronary artery disease of the European Society of Cardiology (ESC) and of the European Association for Cardio-Thoracic Surgery (EACTS). *Eur Heart J*. 2017;39(3):213−260.

41. Levine GN, Bates ER, Bittl JA, et al. 2016 ACC/AHA guideline focused update on duration of dual antiplatelet therapy in patients with coronary artery disease. A report of the American College of Cardiology/American Heart Association Task Force on Clinical Practice Guidelines. *Circulation*. 2016;68(10):1082−1115.

42. Steg PG, Bhatt DL, Simon T, et al. Ticagrelor in patients with stable coronary disease and diabetes. *N Engl J Med*. 2019;381(14):1309−1320.

43. Neumann F-J, Sousa-Uva M, Ahlsson A, et al. 2018 ESC/EACTS Guidelines on myocardial revascularization. *Eur Heart J*. 2018;40(2):87−165.

44. Sibbing D, Aradi D, Jacobshagen C, et al. Guided de-escalation of antiplatelet treatment in patients with acute coronary syndrome undergoing percutaneous coronary intervention (TROPICAL-ACS): a randomised, open-label, multicentre trial. *Lancet*. 2017; 390(10104):1747−1757.

Peripheral arterial disease—a different kind of arterial disease? The role of antiplatelet therapy in the prevention and treatment of limb ischemia

Sudhakar Sattur, MD, MHSA

Guthrie Clinic and Robert Packer Hospital, Sayre, PA, United States

Introduction

Lower extremity peripheral arterial disease (PAD) is a common cardiovascular disease that affects approximately 8.5 million Americans above the age of 40 years and is associated with significant morbidity and mortality.[1] Current guidelines recommend aggressive risk factor modification along with guideline-directed medical therapy in patients with PAD.[2,3] Antiplatelet therapy is one of the mainstays of treatment in these patients. However, the supporting evidence is limited and continues to evolve.

This chapter summarizes the current evidence for the role of antiplatelet therapy in patients with PAD. It is outlined under the following sections:

1. Asymptomatic patients
   a. Patients with the PAD alone, without clinically manifest coronary artery disease (CAD) or cerebrovascular disease.
   b. Patients with PAD and clinically manifest CAD or cerebrovascular disease.
2. Symptomatic patients
3. Revascularization
   a. Endovascular revascularization
   b. Surgical revascularization

Patients with peripheral arterial disease alone without clinically manifest coronary artery or cerebrovascular disease

The American College of Cardiology/American Heart Association (ACC/AHA) guidelines recommend aspirin or clopidogrel therapy to reduce cardiovascular events (class II recommendation). On the other hand, the European Society of Cardiology/European Society of Vascular Surgery (ESC/ESVS) guidelines recommend against antiplatelet therapy in patients with asymptomatic PAD unless they have other indications for antiplatelet therapy (class III recommendation).[2,3] The latter recommendation is based on two clinical trials that evaluated the effect of aspirin on cardiovascular outcomes in patients with asymptomatic PAD: the AAA (Aspirin for Asymptomatic Atherosclerosis) and the POPADAD (Prevention of Progression of Arterial Disease and Diabetes) trials.[4,5] In the AAA trial, 3350 patients with an ankle brachial index (ABI) of ≤0.95 were randomized to aspirin or placebo. After a mean follow-up of 8.2 years, there was no significant difference in major adverse cardiac or cerebrovascular events (MACCE) between the two groups (10.8% for aspirin vs. 10.5% for placebo; hazard ratio [HR]: 1.03 [95% confidence interval (CI), 0.84−1.27]). The POPADAD trial randomized 1276 patients with an ABI of ≤0.99 and diabetes and found no significant difference in the composite risk of MACCE or amputation between aspirin and placebo (18.2% vs. 18.3%, respectively; HR: 0.98 [95% CI, 0.76−1.26). A major limitation of both these trials is the use of noncontemporary definition of PAD (ABI less than 0.99 in the POPADAD trial and ABI less than 0.95 in the AAA trial).

Patients with peripheral arterial disease and clinically manifest coronary artery or cerebrovascular disease

These patients have asymptomatic PAD with a history of clinically manifest CAD or cerebrovascular disease. Antiplatelet therapy with aspirin, P2Y$_{12}$ inhibitors, or both should be recommended according to current guidelines for treating acute (ischemic event within the past 12 months) or stable CAD or cerebrovascular disease.[6−8]

Patients with symptomatic peripheral arterial disease

It is a class IA recommendation in the ACC/AHA guidelines to use antiplatelet therapy with aspirin alone (range, 75−325 mg/day) or clopidogrel alone (75 mg/day) in patients with symptomatic PAD to reduce the risk of myocardial infarction (MI), stroke, or vascular death.[2] In the ESC/ESVS guidelines, long-term single antiplatelet therapy is also a class IA recommendation for patients with symptomatic PAD .[3] However, European guidelines do not specifically recommend aspirin as a single antiplatelet therapy in these patients. Evidence related to aspirin and other agents including P2Y$_{12}$ inhibitors is described in the following.

Aspirin monotherapy

In a meta-analysis of 42 trials comprising 9706 patients with symptomatic PAD, antiplatelet therapy was associated with a 23% reduction in serious vascular events (nonfatal MI, nonfatal stroke, or vascular death), from 7.1% to 5.8%, $P = 0.004$.[9] A subsequent meta-analysis involving 5269 individuals compared the efficacy of aspirin with or without dipyridamole with placebo in both symptomatic and asymptomatic PAD.[10] The authors observed a nonsignificant decrease in MACCE (8.9% for aspirin ± dipyridamole vs. 11.0% for placebo; risk ratio [RR]: 0.88 [95% CI, 0.76−1.04]). The CLIPS (Critical Leg Ischemia Prevention Study) trial randomized 366 patients with symptomatic and asymptomatic PAD to aspirin versus placebo.[11] Aspirin significantly reduced the risk of a composite of MACCE and critical limb ischemia (6.5% for aspirin vs. 15.5% for placebo; HR: 0.42 [95% CI, 0.21−0.82]) at nearly 2-year follow-up.

P2Y$_{12}$ inhibitor monotherapy

Clopidogrel has a superior efficacy and safety profile when used as a single antiplatelet agent in patients with symptomatic PAD. The CAPRIE (Clopidogrel versus Aspirin in Patients at Risk of Ischemic Events) trial compared clopidogrel 75 mg daily with aspirin 325 mg in patients with recent ischemic stroke, recent MI, or symptomatic PAD.[12] Clopidogrel was superior to aspirin in reducing MACCE over a mean follow-up period of 1.91 years (5.3% for clopidogrel vs. 5.8% for aspirin; RR: 0.91 [95% CI, 0.84−0.997]). The greatest risk reduction was seen in the PAD subgroup (3.7% for clopidogrel vs. 4.9% for aspirin; RR: 0.76 [95% CI, 0.64−0.91]). Clopidogrel was associated with a significantly lower number of hospitalizations not only for ischemic but also for bleeding events (1502 vs. 1673; $P = 0.010$).[13] The EUCLID (Examining Use of Ticagrelor in Peripheral Artery Disease) trial compared ticagrelor with clopidogrel in 13,885 patients with PAD.[14] There was no significant difference in MACCE between ticagrelor and clopidogrel (10.8% vs. 10.6%, respectively; HR: 1.02 [95% CI, 0.92−1.13], $P = 0.65$) at 2.5 years of follow-up. Major bleeding rates were also similar between the two groups (1.6% vs.1.6%; HR 1.10; 95% CI, 0.84−1.43; $P = 0.49$). Thus ticagrelor has comparable efficacy and safety to clopidogrel in the treatment of symptomatic PAD.

Dual antiplatelet therapy

The CHARISMA (Clopidogrel for High Atherothrombotic Risk and Ischemic Stabilization, Management, and Avoidance) trial evaluated patients with established cardiovascular disease or multiple atherothrombotic risk factors.[15] In a subanalysis of CHARISMA trial, which included those with a history of stroke, MI, or symptomatic PAD, the rate of MACCE was significantly lower in the aspirin plus clopidogrel arm (7.3% for aspirin plus clopidogrel vs. 8.8% for aspirin; HR: 0.83 [95% CI, 0.72−0.96], $P = 0.01$).[16] In the PAD subgroup, dual antiplatelet therapy (DAPT) reduced the risk of myocardial infraction (HR: 0.63 [95% CI, 0.42−0.96], $P = 0.03$) and hospitalization for ischemic events (HR: 0.81 [95% CI, 0.68−0.95], $P = 0.01$), without significantly affecting the rate of MACCE (HR: 0.85 [95% CI, 0.66−1.08], $P = 0.18$).[17] Rate of severe bleeding (according to the GUSTO [global utilization of streptokinase and tPA for occluded arteries] criteria) was 1.7% in the DAPT group and 1.3% in the aspirin group (RR: 1.25 [95% CI,

0.97−1.61], $P = 0.09$). The rate of moderate bleeding was 2.1% in the DAPT group, as compared with 1.3% in the aspirin group (RR: 1.62 [95% CI, 1.27−2.08], $P < 0.001$).

The PLATO (Platelet Inhibition and Patient Outcomes) trial compared ticagrelor and aspirin with clopidogrel and aspirin in patients with acute coronary syndrome.[18] In a subgroup analysis of the PLATO trial, which included 1144 PAD patients with PAD, the rate of MACCE was insignificantly lower in the ticagrelor and aspirin group compared with that in the clopidogrel and aspirin group (18.0% vs. 20.6%, respectively; HR: 0.85 [95% CI, 0.64−1.11]).[19] Also noteworthy is the markedly higher rate of events in the PAD group compared with non-PAD group, regardless of treatment assignment (19% vs. 10.2%, respectively, $P < 0.001$). In the Prevention of Cardiovascular Events in Patients with Prior Hear Attack Using Ticagrelor Compared to Placebo on a Background of Aspirin-Thrombolysis in Myocardial Infarction (PEGASUS-TIMI) 54 trial, which evaluated patients with a prior MI, ticagrelor and aspirin were compared to aspirin alone. Patients treated for 3 years with ticagrelor and aspirin had significantly reduced risk of MACCE (90-mg group: HR 0.85 [95% CI, 0.75−0.96] and 60-mg group: HR 0.84 [95% CI, 0.74−0.95]).[20] In a prespecified subgroup analysis of 1143 PAD patients, the rates of MACCE and major adverse limb events (MALE) were lower in the ticagrelor and aspirin arm than in the aspirin group (14.1% vs. 19.3%; HR: 0.69 [95% CI, 0.47−0.99, $P = 0.04$] and 0.46% vs. 0.71%; HR: 0.65 [95% CI, 0.44 to 0.95], $P = 0.03$, respectively).[21] Again noted was a 2.5-fold higher rate of events in patients with PAD compared to those without PAD. Thrombolysis in Myocardial Infarction major or minor bleeding was not significantly different in the ticagrelor group compared to the clopidogrel group (ticagrelor 60-mg group: HR 1.36 (0.46−4.05), $P = 0.58$ and ticagrelor 90-mg group: HR 1.57 (0.54−4.53), $P = 0.40$).

Subgroup analysis of patients with PAD from the DAPT (Dual Antiplatelet Therapy) trial, comparing 30 versus 12 months of DAPT after coronary artery stenting, showed an insignificant lower risk of MACCE in the PAD subgroup with prolonged therapy (4.7% vs. 7.3%, respectively; HR: 0.63 [95% CI, 0.32−1.22]).[22] Another trial, PRODIGY (Prolonging Dual Antiplatelet Treatment After Grading Stent-Induced Intimal Hyperplasia Study) compared long and short duration of DAPT (24 vs. 6 months) after coronary stenting.[23] In a prespecified subgroup analysis of patients with PAD, there was a significant reduction in MACCE (16.1% vs. 27.3%; HR: 0.54 [95% CI, 0.31−0.95], $P = 0.03$). Although moderate and severe bleeding rates were significantly increased among DAPT and PRODIGY trial patients treated with longer duration of DAPT, this was not seen in the subgroup analysis of patients with PAD (adjusted HR, 0.75; 95% CI, 0.42−1.33; $P = 0.32$).

Protease-activated receptor 1 antagonist

The efficacy and safety of vorapaxar when used in conjunction with other antiplatelet agents was examined in the TRA2°P-TIMI 50 (Thrombin Receptor Antagonist in Secondary Prevention of Atherothrombotic Ischemic Events-Thrombolysis in

Myocardial Infarction 50) and TRACER (Thrombin Receptor Antagonist for Clinical Event Reduction in Acute Coronary Syndrome) trials.[24,25] In the TRA2°P-TIMI 50 trial, which included patients with a history of MI, ischemic stroke, or PAD, there was a significant reduction in MACCE with vorapaxar compared with placebo (9.3% vs. 10.5%, respectively; HR: 0.87 [95% CI, 0.80−0.94], $P < 0.001$).[26] There was no significant reduction in overall MACEE in the cohort analysis of patients with PAD (11.3% for vorapaxar vs. 11.9% for placebo; HR: 0.94 [95% CI, 0.78−1.14]). However, rates of hospitalization for acute limb ischemia (2.3% vs. 3.9%; HR, 0.58; 95% CI, 0.39−0.86; $P = 0.006$) and peripheral artery revascularization (18.4% vs. 22.2%; HR, 0.84; 95% CI, 0.73−0.97; $P = 0.017$) were significantly lower in patients randomized to vorapaxar.[27] The TRACER trial evaluated the efficacy and safety of vorapaxar in patients with acute coronary syndrome. In the subgroup analysis of 936 PAD patients, there was a nonsignificant reduction in MACCE (21.7% for vorapaxar vs. 24.8% for placebo; HR: 0.85 [95% CI, 0.64−1.13]). The risk of lower extremity vascularization and amputations was also numerically lower in the vorapaxar group. Bleeding was significantly higher in the vorapaxar group in both the trials (TRA2°P-TIMI 50 trial: 7.4% vs. 4.5%, HR 1.62 [1.21−2.18], $P = 0.001$) and TRACER trial: 9.1% vs. 5%, HR 1.47, 95% CI [0.89−2.45], $P = 0.004$).

Anticoagulation

The WAVE (Warfarin Antiplatelet Vascular Evaluation) trial compared the efficacy and safety of warfarin plus antiplatelet therapy with antiplatelet alone in 2161 PAD patients.[28] There were no significant differences in MACCE and severe ischemia of the peripheral or coronary arteries (15.9% vs. 17.4%, respectively; RR: 0.91 [95% CI, 0.74−1.12]) with anticoagulation and antiplatelet therapy compared to antiplatelet therapy alone. There was significantly more life-threating bleeding in the anticoagulation group (4.0% vs. 1.2%; RR: 3.41 [95% CI, 1.84−6.35], $P < 0.001$).

The more recent COMPASS (Cardiovascular Outcomes for People Using Anticoagulation Strategies) trial compared two different doses of rivaroxaban plus aspirin to aspirin alone in patients with stable atherosclerotic cardiovascular disease (rivaroxaban 2.5 mg twice daily plus aspirin 100 mg daily, rivaroxaban 5 mg twice daily, or aspirin 100 mg alone).[29] The PAD cohort analysis of the COMPASS trial showed efficacy of rivaroxaban plus aspirin combination therapy compared with aspirin alone (7470 patients with PAD or carotid artery disease: 5.1% vs. 6.9%, respectively; HR: 0.72 [95% CI, 0.57−0.90], $P = 0.005$).[30] There were fewer MALE in the rivaroxaban plus aspirin group than in the aspirin-alone group (1.2% vs. 2.2%, respectively; HR: 0.54 [95% CI, 0.35−0.84], $P = 0.004$), as well as fewer major amputations (0.2% vs. 0.7%, respectively; HR: 0.30 [95% CI, 0.11−0.80]). Rivaroxaban 5 mg twice a day compared with aspirin alone did not significantly reduce the composite MACCE (149 [6%] of 2474 vs. 174 [7%] of 2504; HR 0.86, 95% CI, 0.69−1.08, $P = 0.19$) but reduced MALE including major amputation (40 [2%] vs. 60 [2%]; HR 0.67, 95% CI, 0.45−1.00, $P = 0.05$). The use of the rivaroxaban plus aspirin combination increased major bleeding compared with the aspirin-alone group (77 [3%] of 2492 vs. 48 [2%] of 2504; HR 1.61, 95% CI, 1.12–2.31, $P = 0.0089$), which was mainly gastrointestinal. The COMPASS trial suggests that rivaroxaban dosed at 2.5 mg twice daily plus aspirin may be a preferred treatment strategy in patients with PAD with lower bleeding risk profile.

Both ACC/AHA and ESC guidelines strongly recommend against the use of therapeutic anticoagulation for patients with stable PAD.[2,3] However, the results of the COMPASS trial were published only after these guideline statements were released. Future updates to these guidelines will incorporate results of COMPASS and other trials.

Endovascular revascularization

High-quality data on the use of antiplatelet therapy after endovascular revascularization for PAD is lacking. A Cochrane meta-analysis evaluated the effect of antiplatelet and anticoagulant drugs in the prevention of restenosis and reocclusion after endovascular treatment.[31] This meta-analysis included 22 trials and a total of 3529 patients. Several different treatment options were studied, including high-dose and low-dose aspirin alone or aspirin plus dipyridamole, clopidogrel, DAPT, vitamin K antagonist, and low-molecular-weight heparin. At 6 months after procedure, treatment with high-dose aspirin plus dipyridamole was associated with a significant reduction in reocclusion, but this was not observed with low-dose aspirin. However, at 12 months after treatment, no significant difference in rates of reocclusion and restenosis was observed among all regimens. Bleeding data was not reported in all trials, but the available evidence suggests greater incidence of gastrointestinal bleeding with high-dose aspirin treatment. The MIRROR (Management of Peripheral Arterial Interventions with Mono or Dual Antiplatelet Therapy) trial randomized 80 patients undergoing peripheral angioplasty with or without stenting to clopidogrel plus aspirin for 6 months versus aspirin alone.[32] Rate of target lesion revascularization was lower in the clopidogrel plus aspirin group than in the aspirin-alone group at 6 months (5% vs. 8%; $P = 0.04$); however, this benefit was lost at 1 year (25% vs. 32%; $P = 0.35$). The STOP-IC (Sufficient Treatment of Peripheral Intervention by Cilostazol) trial randomized 200 patients after undergoing percutaneous transluminal angioplasty with stenting for femoropopliteal lesions to aspirin plus cilostazol or aspirin alone.[33] Cilostazol reduced the rate of restenosis compared to aspirin alone (20% vs. 49%, $P = 0.0001$). Rates of cardiovascular events were similar in the two groups. Data on bleeding events were not reported. The VOYAGER PAD (Vascular outcomes study of aspirin along with Rivaroxaban in endovascular or surgical revascularization for PAD) trial randomly assigned 6564 PAD patients who had undergone lower extremity revascularization to take rivaroxaban 2.5 mg twice daily plus aspirin or placebo and aspirin.[34] Surgical revascularization was the approach in 35% and endovascular or a hybrid approach in 77% patients. There were significantly lower primary event rates at 3-year follow-up (acute limb ischemia, major amputation for vascular causes, MI, ischemic stroke, or cardiovascular death) in the rivaroxaban group compared with the placebo group (17.3% vs. 19.9%, HR: 0.85; 0.76—0.96; $P = 0.009$).There was significantly higher bleeding in the rivaroxaban group (Thrombolysis in Myocardial Infarction major bleeding occurred in 62 patients in the rivaroxaban group vs. 44 in the placebo group, HR: 1.43; 0.97—2.10, $P = 0.07$); however, there was no significant increase in fatal or intracranial bleeding. In a prespecified analysis, half of the patients in the dual therapy group were given clopidogrel for 6 months. There was no significant benefit with triple therapy (16% vs. 18.3% without clopidogrel; HR 0.85, CI, 0.71—1.01). Major bleeding rates were increased with triple therapy compared to dual therapy.

Surgical revascularization

A Cochrane review of 954 patients undergoing bypass graft surgery suggested a benefit of aspirin with or without dipyridamole on infrainguinal bypass graft patency at 1 year (odds ratio: 0.42 [95% CI, 0.22—0.83], $P = 0.01$).[35] The CASPAR (Clopidogrel and Acetylsalicylic acid in bypass Surgery for Peripheral Arterial disease) trial evaluated the efficacy of DAPT with clopidogrel and aspirin versus aspirin alone in 851 patients undergoing below-the-knee bypass grafting.[36] There was no significant difference in the composite rate of graft occlusion, revascularization, amputation, or death between the two arms (35.1% for clopidogrel plus aspirin vs. 35.4% for aspirin; HR: 0.98 [95% CI, 0.78—1.23]) at 1-year follow-up. Bleeding was higher with clopidogrel plus aspirin than that with aspirin alone (16.7% vs. 7.1%, respectively; $P < 0.001$), although no significant differences were observed in the rates of severe bleeding between the two groups (1.2% vs. 2.1%).

The Dutch BOA (Bypass Oral anticoagulants or Aspirin) trial randomized 2690 patients undergoing infrainguinal bypass to high-intensity anticoagulation with vitamin K antagonist versus aspirin.[37] Rates of graft occlusion were similar between the two groups (13.5% for anticoagulation vs. 14.2% for aspirin; HR: 0.95 [95% CI, 0.82—1.11]). Severe bleeding including intracranial hemorrhage was higher with anticoagulation than that with aspirin (8.1% vs. 4.2%, respectively; HR: 1.96 [95% CI, 1.42—2.71]). Sarac et al.[38] randomized patients undergoing infrainguinal vein bypass surgery to warfarin plus aspirin versus aspirin alone. These patients were deemed high risk for bypass graft closure due to suboptimal conduit, poor runoff, or redo procedure. The 3-year primary patency and limb salvage rates were significantly higher with warfarin plus aspirin than with aspirin alone (74% vs. 51%, $P = 0.04$ and 81% vs. 31%, $P = 0.02$, respectively).

The ESC guidelines recommend DAPT for ≥ 1 month after endovascular revascularization. Also, the ESC guidelines recommend considering DAPT after below-the-knee prosthetic bypass and anticoagulation with a vitamin K antagonist after venous bypass.

Plaque and lesion characteristics of peripheral arterial disease versus coronary artery disease

Long-term outcomes of balloon angioplasty, atherectomy, and stent placement in patients with PAD are inferior to similar interventions used for treatment of patients with CAD.[39,40] One of the main reasons for these differing outcomes could be related to differences in plaque characteristics. The PAD plaques have more calcium than CAD plaques. There is more medial calcification with PAD compared with intimal calcification with CAD. There is more fibrosis and less inflammation with PAD plaques compared with CAD plaques. Additionally, the PAD plaques tend to be more concentric and diffuse compared with CAD plaques. These differences may explain the more heterogeneous response to antiplatelet therapy in patients with PAD, compared with the uniform benefit seen in CAD.

Summary

Robust evidence about antiplatelet therapy in PAD is lacking. The majority of the recommendations are derived from the treatment of other vascular distributions, predominantly CAD. However, PAD has several unique characteristics that call for dedicated studies evaluating antiplatelet therapy in these patients. Recent contemporary trials (such as COMPASS and VOYAGER PAD trials) are attempting to fill this void. Large randomized studies evaluating the role of antiplatelet therapy in selected clinical scenarios of PAD are warranted.

References

1. Mozaffarian D, Benjamin EJ, Go AS, et al. Heart disease and stroke statistics—2016 update: a report from the American Heart Association. *Circulation.* 2016;133:e38—e360.
2. Gerhard-Herman MD, Gornik HL, Barrett C, et al. 2016 AHA/ACC guideline on the management of patients with lower extremity peripheral artery disease: executive summary: a report of the American College of Cardiology/American Heart Association Task Force on Clinical Practice Guidelines. *J Am Coll Cardiol.* 2017;69:1465—1508.
3. Aboyans V, Ricco J-B, Bartelink M-LEL, et al. 2017 ESC guidelines on the diagnosis and treatment of peripheral arterial diseases, in collaboration with the European Society for Vascular Surgery (ESVS): document covering atherosclerotic disease of extracranial carotid and vertebral, mesenteric, renal, upper and lower extremity arteries endorsed by: the European Stroke Organization (ESO) the task force for the diagnosis and treatment of peripheral arterial diseases of the European Society of Cardiology (ESC) and of the European Society for Vascular Surgery (ESVS). *Eur Heart J.* 2018;39:783—816.
4. Fowkes FGR, Price JF, Stewart MCW, et al. Aspirin for prevention of cardiovascular events in a general population screened for a low ankle brachial index: a randomized controlled trial. *J Am Med Assoc.* 2010;303:841—848.

5. Belch J, MacCuish A, Campbell I, et al. The prevention of progression of arterial disease and diabetes (POPADAD) trial: factorial randomised placebo controlled trial of aspirin and antioxidants in patients with diabetes and asymptomatic peripheral arterial disease. *Br Med J*. 2008;337:a1840.

6. Powers WJ, Rabinstein AA, Ackerson T, et al. Guidelines for the early management of patients with acute ischemic stroke: 2019 update to the 2018 guidelines for the early management of acute ischemic stroke: a guideline for healthcare professionals from the American Heart Association/American Stroke Association [published correction appears in Stroke. 2019 Dec;50(12):e440−e441] *Stroke*. 2019;50(12):e344−e418.

7. Fihn SD, Blankenship JC, Alexander KP, et al. 2014 ACC/AHA/AATS/PCNA/SCAI/STS focused update of the guideline for the diagnosis and management of patients with stable ischemic heart disease: a report of the American College of Cardiology/American Heart Association Task Force on Practice Guidelines, and the American Association for Thoracic Surgery, Preventive Cardiovascular Nurses Association, Society for Cardiovascular Angiography and Interventions, and Society of Thoracic Surgeons. *J Am Coll Cardiol*. 2014;64(18):1929−1949.

8. Levine GN, Bates ER, Bittl JA, et al. 2016 ACC/AHA guideline focused update on duration of dual antiplatelet therapy in patients with coronary artery disease: a report of the American College of Cardiology/American Heart Association Task Force on Clinical Practice Guidelines: an Update of the 2011 ACCF/AHA/SCAI guideline for percutaneous coronary intervention, 2011 ACCF/AHA guideline for coronary artery bypass graft surgery, 2012 ACC/AHA/ACP/AATS/PCNA/SCAI/STS guideline for the diagnosis and management of patients with stable ischemic heart disease, 2013 ACCF/AHA guideline for the management of ST-elevation myocardial infarction, 2014 AHA/ACC guideline for the management of patients with non-ST-elevation acute coronary syndromes, and 2014 ACC/AHA guideline on perioperative cardiovascular evaluation and management of patients undergoing noncardiac surgery [published correction appears in Circulation. 2016 Sep 6;134(10):e192−4] *Circulation*. 2016;134(10):e123−e155.

9. Antithrombotic Trialists' Collaboration. Collaborative meta-analysis of randomised trials of antiplatelet therapy for prevention of death, myocardial infarction, and stroke in high risk patients. *Br Med J*. 2002;324:71−86.

10. Berger JS, Krantz MJ, Kittelson JM, Hiatt WR. Aspirin for the prevention of cardiovascular events in patients with peripheral artery disease: a meta-analysis of randomized trials. *J Am Med Assoc*. 2009;301:1909−1919.

11. Critical Leg Ischaemia Prevention Study (CLIPS) Group, Catalano M, Born G, Peto R. Prevention of serious vascular events by aspirin amongst patients with peripheral arterial disease: randomized, double-blind trial. *J Intern Med*. 2007;261:276−284.

12. CAPRIE Steering Committee. A randomized, blinded, trial of clopidogrel versus aspirin in patients at risk of ischemic events (CAPRIE). CAPRIE Steering Committee. *Lancet*. 1996;348:1329−1339.

13. Bhatt DL, Hirsch AT, Ringleb PA, Hacke W, Topol EJ. Reduction in the need for hospitalization for recurrent ischemic events and bleeding with clopidogrel instead of aspirin. CAPRIE investigators. *Am Heart J*. 2000;140:67−73.

14. Hiatt WR, Fowkes FGR, Heizer G, et al. Ticagrelor versus clopidogrel in symptomatic peripheral artery disease. *N Engl J Med*. 2017;376:32−40.

15. Bhatt DL, Fox KAA, Hacke W, et al. Clopidogrel and aspirin versus aspirin alone for the prevention of atherothrombotic events. *N Engl J Med*. 2006;354:1706−1717.

16. Bhatt DL, Flather MD, Hacke W, et al. Patients with prior myocardial infarction, stroke, or symptomatic peripheral arterial disease in the CHARISMA trial. *J Am Coll Cardiol.* 2007;49:1982−1988.

17. Cacoub PP, Bhatt DL, Steg PG, Topol EJ, Creager MA, CHARISMA Investigators. Patients with peripheral arterial disease in the CHARISMA trial. *Eur Heart J.* 2009;30: 192−201.

18. Wallentin L, Becker RC, Budaj A, et al. Ticagrelor versus clopidogrel in patients with acute coronary syndromes. *N Engl J Med.* 2009;361:1045−1057.

19. Patel MR, Becker RC, Wojdyla DM, et al. Cardiovascular events in acute coronary syndrome patients with peripheral arterial disease treated with ticagrelor compared with clopidogrel: data from the PLATO trial. *Eur J Prev Cardiol.* 2015;22:734−742.

20. Bonaca MP, Bhatt DL, Cohen M, et al. Long-term use of ticagrelor in patients with prior myocardial infarction. *N Engl J Med.* 2015;372:1791−1800.

21. Bonaca MP, Bhatt DL, Storey RF, et al. Ticagrelor for prevention of ischemic events after myocardial infarction in patients with peripheral artery disease. *J Am Coll Cardiol.* 2016; 67:2719−2728.

22. Secemsky EA, Yeh RW, Kereiakes DJ, et al. Extended duration dual antiplatelet therapy after coronary stenting among patients with peripheral arterial disease: a subanalysis of the dual antiplatelet therapy study. *J Am Coll Cardiol Intv.* 2017;10:942−954.

23. Franzone A, Piccolo R, Gargiulo G, et al. Prolonged vs. short duration of dual antiplatelet therapy after percutaneous coronary intervention in patients with or without peripheral arterial disease: a subgroup analysis of the PRODIGY randomized clinical trial. *JAMA Cardiol.* 2016;1:795−803.

24. Morrow DA, Braunwald E, Bonaca MP, et al. Vorapaxar in the secondary prevention of atherothrombotic events. *N Engl J Med.* 2012;366:1404−1413.

25. Jones WS, Tricoci P, Huang Z, et al. Vorapaxar in patients with peripheral artery disease and acute coronary syndrome: insights from Thrombin Receptor Antagonist for Clinical Event Reduction in Acute Coronary Syndrome (TRACER). *Am Heart J.* 2014;168: 588−596.

26. Bonaca MP, Scirica BM, Creager MA, et al. Vorapaxar in patients with peripheral artery disease: results from TRA2°P-TIMI 50. *Circulation.* 2013;127:1522−1529.

27. Bonaca MP, Gutierrez JA, Creager MA, et al. Acute limb ischemia and outcomes with vorapaxar in patients with peripheral artery disease: results from the trial to assess the effects of vorapaxar in preventing heart attack and stroke in patients with atherosclerosis-thrombolysis in myocardial infarction 50 (TRA2°P-TIMI 50). *Circulation.* 2016;133: 997−1005.

28. Warfarin Antiplatelet Vascular Evaluation Trial Investigators. Oral anticoagulant and antiplatelet therapy and peripheral arterial disease. *N Engl J Med.* 2007;357:217−227.

29. Eikelboom JW, Connolly SJ, Bosch J, et al. Rivaroxaban with or without aspirin in stable cardiovascular disease. *N Engl J Med.* 2017;377:1319−1330.

30. Anand SS, Bosch J, Eikelboom JW, et al. Rivaroxaban with or without aspirin in patients with stable peripheral or carotid artery disease: an international, randomized, double-blind, placebo-controlled trial. *Lancet.* 2018;391:219−229.

31. Robertson L, Ghouri MA, Kovacs F, et al. Antiplatelet and anticoagulant drugs for prevention of restenosis/reocclusion following peripheral endovascular treatment. *Cochrane Database Syst Rev.* 2012;8:CD002071.

32. Tepe G, Bantleon R, Brechtel K, et al. Management of peripheral arterial interventions with mono or dual antiplatelet therapy—the MIRROR study: a randomised and double-blinded clinical trial. *Eur Radiol*. 2012;22:1998–2006.

33. Iida O, Yokoi H, Soga Y, et al. Cilostazol reduces angiographic restenosis after endovascular therapy for femoropopliteal lesions in the Sufficient treatment of peripheral intervention by Cilostazol study. *Circulation*. 2013;127:2307–2315.

34. Bonaca MP, Bauersachs RM, Anand SS, et al. Rivaroxaban in peripheral artery disease after revascularization. *N Engl J Med*. 2020;382(21):1994–2004.

35. Bedenis R, Lethaby A, Maxwell H, Acosta S, Prins MH. Antiplatelet agents for preventing thrombosis after peripheral arterial bypass surgery. *Cochrane Database Syst Rev*. 2015;(2):CD000535. Google Scholar.

36. Belch JJF, Dormandy J, CASPAR Writing Committee, et al. Results of the randomized, placebo-controlled clopidogrel and acetylsalicylic acid in bypass surgery for peripheral arterial disease (CASPAR) trial. *J Vasc Surg*. 2010;52:825–833.

37. Efficacy of oral anticoagulants compared with aspirin after infrainguinal bypass surgery (the Dutch bypass oral anticoagulants or aspirin study): a randomised trial. *Lancet*. 2000; 355:346–351.

38. Sarac TP, Huber TS, Back MR, et al. Warfarin improves the outcome of infrainguinal vein bypass grafting at high risk for failure. *J Vasc Surg*. 1998;28:446–457.

39. Yin D, Matsumura M, Rundback J, et al. Comparison of plaque morphology between peripheral and coronary artery disease (from the CLARITY and ADAPT-DES IVUS substudies). *Coron Artery Dis*. 2017;28(5):369–375.

40. Di Vito L, Silenzi S. Distinctive atherosclerotic features in coronary and peripheral arteries. *Coron Artery Dis*. 2017;28(5):364–365.

Cerebrovascular disease—what is the role of dual antiplatelet therapy for the prevention and treatment of ischemic stroke?

Antonio Greco, MD [1], Davide Capodanno, MD, PhD [2]

[1]*Division of Cardiology, A.O.U. Policlinico "G. Rodolico - San Marco", University of Catania, Catania, Italy;* [2]*Professor of Cardiovascular Diseases, Division of Cardiology, A.O.U. Policlinico "G. Rodolico - San Marco", University of Catania, Catania, Italy*

Introduction

Cerebrovascular disease encompasses a broad spectrum of disorders mainly including stroke and transient ischemic attack (TIA), which are acute events due to a critical loss of cerebral blood flow leading to unstable symptoms and highly impacting morbidity, mortality, and functional disability.[1] Primary prevention aims to avert the onset of stroke and TIA by acting on underlying atherosclerotic risk factors, such as hypertension, smoking, diabetes mellitus, and hyperlipidemia.

After a first-time non-fatal stroke, the risk of recurrent events is substantial, which requires an aggressive secondary prevention strategy, where antithrombotic therapy holds an important place.[2] Platelet activation plays a central role in vascular atherosclerosis and in the pathophysiology of ischemic and noncardioembolic stroke. As such, antiplatelet drugs are the standard of care for stroke medical treatment and prevention of recurrences.[3] Yet, despite their undisputed benefits, antiplatelet drugs convey an increased tendency toward infarct hemorrhagic transformation and systemic bleeding. Therefore, careful balancing individual thrombotic and bleeding risks is necessary, and the optimal antithrombotic regimen and duration for patients with cerebrovascular disease still represent areas of uncertainty.

Mechanisms of stroke

The Trial of Org 10172 in Acute Stroke Treatment (TOAST) categorized ischemic stroke into several subtypes (Fig. 6.1):[4]

1. Large-artery atherosclerosis (20%): Occlusion or stenosis ≥50% in a large intracranial (anterior cerebral, middle cerebral, vertebral, or basilar) or extra-cranial (carotid or vertebral) artery due to atheromatous plaques. This group also includes aortic arch atheroma, which is known to be associated with stroke if large plaques (>4 mm) are present.
2. Cardiogenic embolism (30%): An embolus originates in the heart and migrates to the cerebral arteries causing ischemic stroke. It can be due to heart valvular disease, blood stasis, arrhythmias, or venous thromboembolism associated with a patent foramen ovale.
3. Small-vessel disease (25%): The microatheromatous occlusion of penetrating arteries supplying the deep structures of the brain may cause lacunar stroke, shown on neuroimaging as small (<20 mm) subcortical necrosis in the area supplied by a single vessel.
4. Stroke of other determined causes (5%): Non-atherosclerotic vascular diseases (arterial dissection, vasculitis, vasospasm), hypercoagulable conditions, hematologic disorders, malignancy, and cerebral venous thrombosis.
5. Cryptogenic stroke (25%): After a careful exclusion of all other causes, the stroke source cannot still be determined. More recently, some events of this group have been categorized as embolic stroke of undetermined source.[5]

Guidelines and current practice

Main current societal guidelines (Table 6.1) agree on the efficacy and safety of antiplatelet monotherapy following short-term dual antiplatelet therapy (DAPT) in the treatment and secondary prevention of ischemic stroke.[6–9] In current practice, clinicians sometimes use cilostazol or dipyridamole on top of acetylsalicylic acid (ASA) as alternatives. However, DAPT is often underused because it is considered too risky, especially after a major stroke, susceptible to hemorrhagic transformation. In the past few years, new trials investigated the role of DAPT in this setting and will inform future guidelines.

Primary prevention

The role of ASA in stroke primary prevention has been explored by several random-ized controlled trials (RCTs) and their meta-analyses; overall, in line with recent data on primary prevention of cardiovascular diseases, ASA was not shown to reduce the risk of stroke, while increasing bleeding.[10–13]

Table 6.1 Current societal guidelines on antithrombotic strategies for therapy and prevention of non-cardioembolic ischemic stroke.

Trial (NCT)	Recommendation
AHA/ASA 2018	Aspirin is recommended in patients with acute ischemic stroke within 24–48 hours after onset (COR I, LOE A)
	In patients presenting with minor stroke, treatment for 21 days with DAPT (aspirin and clopidogrel) begun within 24 h can be beneficial for early secondary stroke prevention for a period of up to 90 days from symptom onset (COR IIa, LOE B-R).
	Ticagrelor is not recommended (over aspirin) in the acute treatment of patients with minor stroke (COR III, LOE B-R).
Canadian Stroke Best Practice Guideline 2017	Acetylsalicylic acid (80–325 mg daily), combined acetylsalicylic acid (25 mg) and extended-release dipyridamole (25 mg/200 mg twice daily), or clopidogrel (75 mg daily) are all appropriate options and selection should depend on the clinical circumstances (LOE A).
	Short-term concurrent use of acetylsalicylic acid and clopidogrel (up to 21 days) has not shown an increased risk of bleeding and may be protective following minor stroke or transient ischemic attack (LOE B).
	Long-term use of acetylsalicylic acid and clopidogrel is not recommended for secondary stroke prevention, unless there is an alternate indication (e.g., coronary drug-eluting stent requiring dual antiplatelet therapy), due to an increased risk of bleeding and mortality (LOE A).
	Expert opinion suggests that if a patient experiences stroke while on acetylsalicylic acid, it may be reasonable to consider switching to clopidogrel; if a patient experiences a stroke while on clopidogrel, it may be reasonable to consider switching to combined acetylsalicylic acid (25 mg) and extended-release dipyridamole (200 mg) (LOE C).
Chinese Guidelines for Secondary Prevention of Ischemic Stroke and TIA 2014	Antiplatelet treatment rather than anticoagulation treatment is recommended to patients with non-cardioembolic ischemic stroke or TIA for the purpose of preventing another stroke or cardiovascular events (Grade I recommendation, Class A evidence).

Continued

Table 6.1 Current societal guidelines on antithrombotic strategies for therapy and prevention of non-cardioembolic ischemic stroke.—*cont'd*

Trial (NCT)	Recommendation
	The optimal dosage of aspirin is between 75 and 150 mg. Both aspirin (25 mg) plus dipyridamole (200 mg, twice per day) and aspirin (25 mg) plus cilostazol (100 mg twice per day) are alternative treatments to aspirin or clopidogrel alone.
	Using a combination of aspirin and clopidogrel for 21 days is recommended to patients with minor stroke or high-risk TIA within 24 h of onset (Grade I recommendation, Class A evidence). After 21 days, either aspirin or clopidogrel can be continued for long-term use (Grade I recommendation, Class A evidence).
	For patients who have had ischemic stroke or TIA with severe intracranial arterial stenosis (stenosis of 70%—99%), a combination of aspirin and clopidogrel is recommended for 90 days (Grade II recommendation, Class B evidence). After 90 days, either aspirin or clopidogrel could be continued for long-term use (Grade I recommendation, Class A evidence).
	For patients who have had ischemic stroke or TIA with atherosclerosis of aortic arch, antiplatelet drugs and statins are recommended (Grade II recommendation, Class B evidence). The effect of anticoagulant drugs or a combination of aspirin and clopidogrel is unclear (Grade II recommendation, Class B evidence).
	For patients who have had non-cardioembolic ischemic stroke or TIA, long-term use of combined aspirin and clopidogrel for a long time is not recommended (Grade I recommendation, Class A evidence).
European Guidelines on cardiovascular disease prevention 2016	In patients who have had non-cardioembolic ischemic stroke or TIA, prevention with aspirin only, dipyridamole plus aspirin, or clopidogrel alone is recommended (COR I, LOE A).
	In patients who have had non-cardioembolic cerebral ischemic events, anticoagulation is not recommended (COR III, LOE B).

AHA, *American Heart Association;* ASA, *American Stroke Association;* COR, *Class of Recommendation;* DAPT, *Dual Antiplatelet Therapy;* LOE, *Level of Evidence;* NCT, *National Clinical Trial;* TIA, *Transient Ischemic Attack.*

Similarly, DAPT has no role in this setting. As such, modifiable risk factors should be the main target of primary prevention, while antithrombotic therapy has a role in stroke treatment and secondary prevention.

Treatment and secondary prevention

Since the mechanism underlying stroke may be non-uniform (thrombosis, embolism, hypoperfusion, or less frequent causes), antithrombotic therapy should be tailored accordingly (Fig. 6.1).[14] Although thrombolysis or endovascular thrombectomy may be sometimes essential in the acute phase, antithrombotic therapy is the key in the aftermath of stroke to minimize its disabling consequences and prevent recurrent events. Anticoagulant therapy is the treatment of choice for cardioembolic stroke and conditions associated with thrombophilia, while the optimal therapy for other causes is still a matter of debate. Notably, similar to other vascular diseases settings, the presence of atherosclerosis or endothelial damage suggests the opportunity for antiplatelet therapy. Major questions regarding antiplatelet therapy following stroke deal with the choice of the drug(s), the antiplatelet regimen (e.g., number of antiplatelet medications), and the optimal duration of antiplatelet therapy.

Choice of antiplatelet drugs

Antiplatelet drugs currently indicated for stroke treatment and secondary prevention are ASA, clopidogrel, dipyridamole, and cilostazol.

ASA irreversibly inhibits cyclooxygenases, hindering the production of prostaglandins and thromboxane, particularly thromboxane A_2, thus preventing platelet aggregation. Low-dose ASA has been used for stroke prevention and intracranial atherosclerosis because of its efficacy, safety, and cost-effectiveness.[15] When a stroke occurs while on aspirin, switching to, or adding, an alternative antiplatelet agent should be considered.

Clopidogrel acts by binding irreversibly to platelets' adenosine diphosphate receptor $P2Y_{12}$, thus preventing adenosine diphosphate-mediated platelet activation. Clopidogrel has never been compared to placebo for secondary stroke prevention but is often administered to patients who do not tolerate ASA for non-cardioembolic stroke prevention. On the other hand, the opportunity to combine ASA and clopidogrel has been widely explored, and this DAPT regimen is now considered a pillar of treatment for the initial phase of acute ischemic stroke.[16] A loading dose of clopidogrel may be used to more rapidly reach an efficacious inhibition of platelet aggregation.[17] The POINT (Platelet-Oriented Inhibition in New TIA and Minor Ischemic Stroke) trial, which used a clopidogrel loading dose of 600 mg, showed a mild increase in bleeding; as such, a loading dose of 300 mg (strategy previously tested in the CHANCE [Clopidogrel in High-risk Patients With Acute Non-disabling Cerebrovascular Events] trial) should be preferable.[18]

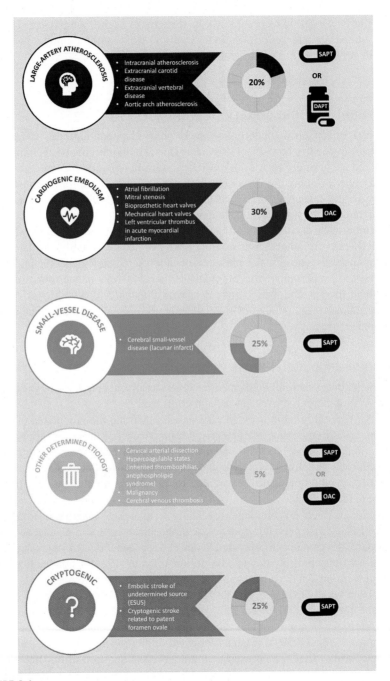

FIGURE 6.1

Management of ischemic stroke subtypes. *DAPT*, Dual Antiplatelet Therapy; *ESUS*, Embolic Stroke of Undetermined Source; *OAC*, Oral Anticoagulation; *SAPT*, Single Antiplatelet Therapy.

At variance with acute coronary syndromes and percutaneous coronary intervention settings, where ASA is typically combined with a $P2Y_{12}$ inhibitor, the combination of ASA and dipyridamole is another DAPT option for stroke treatment and prevention. Dipyridamole is a phosphodiesterase inhibitor and several RCTs showed that its combination with ASA significantly reduces stroke recurrences in comparison with placebo or antiplatelet monotherapy with either ASA or clopidogrel.[19–21]

Cilostazol, a phosphodiesterase-3 inhibitor with additional antiplatelet and vasodilatory effects, has been shown to prevent stroke recurrences better than placebo and similar to ASA.[22] The combination of cilostazol and ASA or clopidogrel has been shown to reduce the incidence of ischemic stroke recurrences among high-risk patients as compared to ASA or clopidogrel alone, without any significant increase in the rate of bleeding complications.[23]

Ticagrelor, a reversible $P2Y_{12}$ inhibitor, was superior to clopidogrel in patients with acute coronary syndrome and a prior stroke or TIA from the PLATO (A Comparison of Ticagrelor [AZD6140] and Clopidogrel in Patients With Acute Coronary Syndrome) trial.[24] However, in the SOCRATES (Acute Stroke Or Transient IsChaemic Attack TReated With Aspirin or Ticagrelor and Patient OutcomES) trial, ticagrelor alone was not better than ASA alone in reducing the composite of death, stroke, or myocardial infarction (hazard ratio [HR] = 0.89, 95% confidence interval [CI], 0.78–1.01, $P = .07$) in patients with acute ischemic stroke, TIA, or symptomatic intracranial or extracranial stenosis; however, there was a reduction in the rate of stroke with ticagrelor (all strokes: 5.9% vs. 6.8%, $P = .03$; ischemic stroke: 5.8% vs. 6.7%, $P = .046$), which may support further investigation on the field.[25]

Prasugrel, a non-reversible $P2Y_{12}$ inhibitor, is not recommended in patients who experienced a previous stroke/TIA due to evidence of harm (56% increase in the net composite outcome of ischemia and bleeding) in this subgroup.[26]

Antiplatelet regimens

Single antiplatelet therapy with ASA has long been the mainstay of treatment and secondary prevention of non-cardioembolic stroke. In the past decades, the practice of combining antiplatelet drugs with different mechanisms of action emerged, leading to substantial improvement in patients' outcomes. Nowadays, the evidence supporting DAPT is growing (Table 6.2) and this regimen represents an invaluable strategy.[6] In more than 20 RCTs performed, the results were controversial, with more trials showing benefit for DAPT, rather than equivalence. In the chronic setting, a 3-month course of DAPT has been suggested in patients with known intracranial arterial stenosis, based on the results of the SAMMPRIS (Stenting versus Aggressive Medical Management for Preventing Recurrent Stroke in Intracranial Stenosis) trial.[27] In addition, several trials evaluated the role of DAPT immediately after ischemic stroke or TIA, with mixed results (Table 6.2).[28–31] The CHANCE[32] and POINT[33] are contemporary large randomized multicenter, double-blinded clinical trials comparing DAPT to ASA monotherapy in patients with acute minor stroke or high-risk TIA. DAPT reduced the recurrence of ischemic events in both

CHANCE (HR, 0.68; 95% CI, 0.57–0.81; $P < .001$) and POINT (HR, 0.75; 95% CI, 0.59–0.95; $P = .02$), while significantly increasing major bleeding in POINT (HR, 2.32; 95% CI, 1.10–4.87; $P = .02$) but not in CHANCE (HR, 1.41; 95% CI, 0.95–2.10; $P = .09$). Some key differences in the trial design of CHANCE and POINT should be mentioned: (1) CHANCE enrolled only Asian patients (known to have different response to antithrombotic agents), whereas POINT was performed in America and Europe; (2) POINT patients underwent randomization earlier (within 12 h of symptoms onset) than CHANCE patients (within 24 h); (3) the clopidogrel loading dose was 600 mg in POINT and 300 mg in CHANCE, and DAPT durations were 90 and 21 days, respectively; (4) bleeding definitions significantly differ between the trials; and (5) the primary endpoint was stroke (ischemic and hemorrhagic) in CHANCE, whereas POINT used a composite of ischemic stroke, myocardial infarction, or death from an ischemic vascular event.

Importantly, a number of meta-analyses report that immediately after an acute ischemic and non-cardioembolic stroke, the absolute number of major ischemic events prevented by DAPT exceeds the provoked increase in major bleeding.[34–36] Among 29,032 patients included in 16 RCTs, compared with monotherapy, DAPT was associated with significantly lower rates of any stroke (risk ratio [RR] 0.80; 95% CI, 0.72–0.89) and ischemic stroke (RR 0.75; 95% CI, 0.66–0.85) during any follow-up period. Although DAPT significantly increased intracranial bleeding (RR 1.55; 95% CI, 1.20–2.01) and major bleeding (RR 1.90; 95% CI, 1.33–2.72), especially with long-term follow-up, the number needed to harm was high (258 and 113, respectively). Short-duration DAPT ($\leq$1 month) started during the early acute ischemic phase was associated with less bleeding than longer duration DAPT and greater reduction in recurrent strokes compared with monotherapy. Other clinical outcomes were essentially similar between the two groups and included recurrent TIA (RR 0.88; 95% CI, 0.72–1.07), myocardial infarction (RR 1.04; 95% CI, 0.84–1.29), vascular death (RR 0.99; 95% CI, 0.82–1.19), and any death (RR 1.12; 95% CI, 0.88–1.42).[35] A meta-regression analysis reported a positive correlation between DAPT efficacy and the severity of stroke, which is an independent predictor of severe outcome; other predictors of recurrent events and better response to DAPT were older age, history of prior stroke/TIA, hypertension, and concomitant multisite atherosclerosis. Conversely, diabetes mellitus reduced the efficacy of DAPT, probably due to higher baseline platelet reactivity and impaired response to clopidogrel.[37] Indeed, in the PRINCE (Effect of Ticagrelor compared with Clopidogrel on High On-treatment Platelet Reactivity in Acute Stroke or Transient Ischemic Attack) trial, conducted in patients with acute stroke or TIA, a DAPT regimen with ASA and ticagrelor reduced the 90-day proportion of patients with high on-treatment platelet reactivity compared to a DAPT regimen with ASA and clopidogrel (RR 0.40; 95% CI, 0.28–0.56; $P < .001$).[38] Furthermore, the recent THALES (Acute STroke or Transient IscHaemic Attack Treated With TicAgreLor and ASA for PrEvention of Stroke and Death) trial randomized 11,016 patients with non-severe non-cardioembolic ischemic stroke to ticagrelor or placebo on top of ASA. DAPT with ASA + ticagrelor reduced the risk of the composite of stroke or death at 30 days, while the incidence of disability was similar between the two groups and severe bleeding was more frequent in the DAPT group.[39]

Other antithrombotic strategies are constantly under evaluation. For instance, the TARDIS (Triple Antiplatelets for Reducing Dependency after Ischaemic Stroke) trial compared a triple antiplatelet regimen (ASA + clopidogrel + dipyridamole) to clopidogrel alone or to the combination of ASA and dipyridamole among patients with acute cerebral ischemia, finding that the more intensive antiplatelet regimen did not reduce the incidence and severity of recurrent stroke or TIA (adjusted odds ratio, 0.90; 95% CI, 0.67−1.20; $P = .47$), while significantly increasing the risk of major bleeding (adjusted common odds ratio, 2.54; 95% CI, 2.05−3.16; $P < .0001$).[40]

Meanwhile, the COMPASS (Cardiovascular Outcomes for People Using Anticoagulation Strategies) trial found that low-dose rivaroxaban on top of ASA is associated with lower rates of ischemic stroke and other vascular outcomes at long-term follow-up compared to antiplatelet monotherapy, without a simultaneous increase in intracranial or fatal bleeding events.[41] Further studies on combined strategies are needed in order to better assess their safety in different patient subsets.

Antiplatelet therapy duration

Although the importance of DAPT in stroke patients is established, uncertainties remain regarding its optimal duration. A pooled analysis of the CHANCE and POINT trials showed that the net benefit of DAPT (prevented ischemic events relative to provoked bleedings) is more evident within the first 3 weeks and even more pronounced in the first 10 days of treatment.[34] In the first 21 days, when DAPT with ASA and clopidogrel was administered in both DAPT trial arms, DAPT reduced the incidence of major ischemic events compared to antiplatelet monotherapy (adjusted HR, 0.66; 95% CI, 0.56−0.77; $P < .001$). From day 22 to day 90, when CHANCE prescribed clopidogrel alone and POINT still prescribed DAPT with ASA and clopidogrel, there were no significant between-group differences in the rates of major ischemic events (adjusted HR, 0.94; 95% CI, 0.67−1.32; $P = .72$), due to a nonsignificant reduction in CHANCE and a nonsignificant increase in POINT. Furthermore, DAPT numerically increased the risk of major bleeding both in the first 21 days (adjusted HR, 2.11; 95% CI, 0.86−5.17; $P = .10$) and from day 22 to day 90 (adjusted HR, 1.28; 95% CI, 0.58−2.81; $P = .55$). Overall, DAPT showed a net benefit both in the total period of 90 days and within the first 21 days, while the partial net benefit from day 22 to day 90 was not significant. The abovementioned findings suggest that DAPT should be routinely administered to acute stroke patients within 24 h of symptom onset and for 10−21 days, followed by single antiplatelet therapy.[6−8]

Table 6.2 Randomized trials on DAPT for ischemic stroke treatment and prevention.

Trial (year)	Blinding	Country	Size	Population	Intervention arm	Control arm	Therapy onset	Primary endpoint	Follow-up	Results
Matias-Guiu et al.[42] (1987)	Open label	Spain	243	Reversible ischemic attack	Dipyridamole (100 mg thrice daily) + ASA (50 mg OD)	Dipyridamole (100 mg four times daily)	≤72 h	Stroke or death	21 months (mean)	Adding ASA to dipyridamole did not convey significantly beneficial effects
ESPS-1[43] (1987)	Double	Europe	2,500	Atherothrombotic stroke, TIA, or reversible ischemic neurologic deficit	ASA (325 mg thrice daily) + dipyridamole (75 mg thrice daily)	Placebo	NA	Stroke or all-cause death	24 months	The primary endpoint occurred less frequently in combination group
ESPS-2[19] (1996)	Double	Europe	6,602	Prior stroke or TIA	ASA (25 mg twice daily) or dipyridamole (200 mg twice daily) or their combination	Placebo	≤72 h	Stroke, death, or their composite	24 months	ASA and dipyridamole are equally effective, while their combination is significantly more effective than either agent alone
MATCH[44] (2004)	Double	Worldwide	7,559	High-risk patients with recent ischemic stroke or TIA and at least one additional vascular risk factor already on clopidogrel	ASA (75 mg OD) + clopidogrel (75 mg OD)	Clopidogrel (75 mg OD)	≤72 h	Composite of ischemic stroke, MI, vascular death, or rehospitalization for acute ischemia	18 months	DAPT increased the risk of life-threatening or major bleeding and did not reduce major vascular events
Chairangsarit et al.[45] (2005)	NA	Thailand	38	Ischemic stroke	ASA (300 mg OD) + dipyridamole (75 mg thrice a day)	ASA (300 mg OD)	≤48 h	Recurrent ischemic stroke, TIA, and vascular death	6 months	No patient developed endpoint events in both groups
CARESS[28] (2005)	Double	Europe	107	Symptomatic (TIA or stroke) carotid stenosis (≥50%) with asymptomatic microembolic signals	ASA (75 mg OD) + clopidogrel (LD 300 mg, then 75 mg OD)	ASA (75 mg OD)	≤30 months	Asymptomatic microembolic signals on transcranial doppler	7 days	DAPT is more effective than ASA alone in reducing asymptomatic embolization

Trial	Blinding	Region	N	Population	Intervention	Comparator	Time window	Outcome	Follow-up	Results
CHARISMA[29] (2006)	Double	Worldwide	15,603	Patients with clinically evident cardiovascular disease or multiple risk factors	ASA (75–162 mg OD) + clopidogrel (75 mg OD)	ASA (75–162 mg OD)	≤24 h	Composite of MI, stroke, or cardiovascular death	28 months (median)	DAPT was not significantly more effective than ASA alone in reducing the rate of MI, stroke, or cardiovascular death
ESPRIT[20] (2007)	Open-label	Worldwide	2,739	Minor stroke or TIA of presumed arterial origin	ASA (30–325 mg OD) + dipyridamole (200 mg twice daily)	ASA (30–325 mg OD)	≤72 h	Composite of death from vascular causes, nonfatal stroke, nonfatal MI, or major bleeding	3.5 years (mean)	The combination of ASA and dipyridamole reduced the incidence of primary endpoint compared to ASA alone
FASTER[30] (2007)	Double	North America	392	Minor stroke or TIA	ASA (LD 162 mg, then 81 mg OD) + clopidogrel (LD 300 mg, then 75 mg OD)	ASA (LD 162 mg, then 81 mg OD)	≤24 h	Ischemic and hemorrhagic stroke	90 days	No statistically significant results due to premature stop
ACTIVE A[46] (2009)	Double	Worldwide	7,554	Patients with AF at increased risk of stroke and in whom VKA therapy was unsuitable	ASA (75–100 mg OD) + clopidogrel (75 mg OD)	ASA (75–100 mg OD)	NA	Composite of stroke, MI, non-CNS systemic embolism, or death from vascular causes	3.6 years (median)	DAPT reduced major vascular events and increased major hemorrhage with respect to ASA
PRoFESS[21] (2008)	Double	Worldwide	1,360	Ischemic stroke	ASA (25 mg twice daily) + dipyridamole (200 mg twice daily)	Clopidogrel (75 mg OD)	≤72 h	Functional outcome (disability and cognitive function)	90 days	No between-group difference

Continued

Table 6.2 Randomized trials on DAPT for ischemic stroke treatment and prevention.—cont'd

Trial (year)	Blinding	Country	Size	Population	Intervention arm	Control arm	Therapy onset	Primary endpoint	Follow-up	Results
CLAIR[47] (2010)	Open-label	Asia	100	Patients who have had acute ischemic stroke or TIA with symptomatic large-artery stenosis in the cerebral or carotid arteries and microembolic signals on transcranial doppler	ASA (75–160 mg OD) + clopidogrel (LD 300 mg, then 75 mg OD)	ASA (75–160 mg OD)	≤7 days	Proportion of patients who had microembolic signals on transcranial doppler	7 days	DAPT reduced the finding of microembolic signals on transcranial doppler in comparison with ASA alone
TOSS II[48] (2011)	Double	Asia	457	Acute symptomatic stenosis in the M1 segment of middle cerebral artery or the basilar artery	ASA (75–150 mg OD) + cilostazol (100 mg twice daily)	ASA (75–150 mg OD) + clopidogrel (75 mg OD)	14 days	Progression of intracranial atherosclerotic stenosis on MRI angiogram	7 months	No between-group difference detected
SPS3[31] (2012)	Double	America and Spain	3,020	Symptomatic lacunar infarcts identified by MRI	ASA (325 mg OD) + clopidogrel (75 mg OD)	ASA (325 mg OD)	<180 days	Any recurrent stroke (ischemic stroke and intracranial hemorrhage)	3.4 years (mean)	DAPT did not reduce the risk of recurrent stroke and increased the risk of bleeding and death
Nakamura et al.[49] (2012)	NA	Japan	76	Non-cardioembolic ischemic stroke	ASA (300 mg for 4 days, then 100 mg OD) + cilostazol (100 mg twice daily)	ASA (300 mg for 4 days, then 100 mg OD)	≤48 h	Neurologic deterioration or stroke recurrence	6 months	ASA plus cilostazol reduced neurologic deterioration and stroke recurrence compared to ASA alone
CHANCE[32] (2013)	Double	China	5,170	Minor ischemic stroke or high-risk TIA	ASA (75 mg OD) + clopidogrel (LD 300 mg, then 75 mg OD) for 21 days, then clopidogrel alone (75 mg OD)	ASA (75 mg OD)	≤24 h	Ischemic or hemorrhagic stroke	90 days	DAPT is superior to ASA alone in reducing the risk of stroke and does not increase the risk of hemorrhage

Study	Design	Country	Population	Treatment	ASA	Time	Outcome	Follow-up	Conclusion
ECLIPse[50] (2013)	Double	Korea	Patients ≥45 years with first ever lacunar infarction	ASA (100 mg OD) + cilostazol (100 mg twice daily)	ASA (100 mg OD)	7 days	Changes of middle cerebral artery and basilar artery pulsatility indexes on transcranial Doppler	90 days	Adding cilostazol to ASA reduced the pulsatility indexes on transcranial Doppler
Yi et al.[51] (2014)	Open-label	China	Large-artery atherosclerosis stroke	ASA (200 mg OD) + clopidogrel (75 mg OD) for 30 days, then clopidogrel alone (75 mg OD)	ASA (200 mg OD for 30 days, then 100 mg OD)	≤48 h	Recurrent ischemic stroke, neurologic deterioration, peripheral vascular events, and MI	1 month	Neurologic deterioration and recurrent ischemic stroke were lower in patients in the combination therapy group
He et al.[52] (2015)	Open-label	China	Patients aged ≥40 years with minor stroke or TIA	ASA (100 mg OD) + clopidogrel (LD 300 mg, then 75 mg OD)	ASA 300 mg OD	≤72 h	Neurologic deterioration, new/recurrent stroke	14 days	DAPT reduced early neurologic deterioration compared to SAPT
COMPRESS[53] (2016)	Double	Korea	Ischemic stroke of presumed large-artery atherosclerosis origin	ASA (LD 300 mg, then 100 mg OD) + clopidogrel (75 mg OD)	ASA (LD 300 mg, then 100 mg OD)	≤48 h	New symptomatic or asymptomatic ischemic lesion on MRI	30 days	No between-group difference in primary endpoint
POINT[33] (2018)	Double	Worldwide	Minor ischemic stroke or high-risk TIA	ASA (50–325 mg OD) + clopidogrel (LD 600 mg, then 75 mg OD)	ASA (50–325 mg OD)	≤12 h	Composite of major ischemic events (ischemic stroke, MI, or death from ischemic vascular events)	90 days	DAPT lowered the risk of major ischemic events but heightened the risk of major hemorrhage compared to ASA alone

Continued

Table 6.2 Randomized trials on DAPT for ischemic stroke treatment and prevention.—cont'd

Trial (year)	Blinding	Country	Size	Population	Intervention arm	Control arm	Therapy onset	Primary endpoint	Follow-up	Results
CSPS.com[23] (2019)	Open-labe	Japan	1,884	High-risk noncardioembolic ischemic stroke on MRI	Cilostazol (100 mg twice daily) + either ASA (81–100 mg OD) or clopidogrel (50–75 mg OD)	ASA (81–100 mg OD) or clopidogrel (50–75 mg OD)	8–180 days	First recurrence of symptomatic ischemic stroke	6 months	Combination therapy reduced the incidence of ischemic stroke recurrence without an increase in bleeding
Aoki et al.[54] (2019)	Open-label	Japan	1,201	Non-cardioembolic stroke	ASA (81–200 mg OD) + cilostazol (100 mg twice daily)	ASA (81–200 mg OD)	≤48 h	Composite of neurologic deterioration, symptomatic stroke recurrence, or TIA	14 days	Combination therapy did not reduce the rate of short-term neurologic worsening compared to ASA alone
PRINCE[38] (2019)	Open-label	China	675	Minor stroke or TIA	ASA (100 mg OD) for 21 days + ticagrelor (LD 180 mg, then 90 mg twice daily)	ASA (100 mg OD) + clopidogrel (LD 300 mg, then 75 mg OD)	≤24 h	Proportion of patients with high platelet reactivity	90 days	Ticagrelor-based strategy showed lower proportion of patients with residual on-treatment high platelet reactivity
THALES (2020)[39]	Double	Worldwide	11,016	Mild-to-moderate acute non-cardioembolic ischemic stroke	ASA (LD 300–325 mg, 75–100 mg OD) + ticagrelor (LD 180 mg, 90 mg twice daily)	ASA (LD 300–325 mg, 75–100 mg OD)	≤ 24 h	Composite of stroke or death	30 days	DAPT reduced the risk of the composite of stroke or death within 30 days as compared to ASA alone, but increased severe bleeding

ACTIVE A, Atrial Fibrillation Clopidogrel Trial With Irbesartan for Prevention of Vascular Events; AF, Atrial Fibrillation; ASA, Acetylsalicylic Acid; CARESS, Clopidogrel and Aspirin for Reduction of Emboli in Symptomatic Carotid Stenosis; CHANCE, Clopidogrel in High-risk Patients With Acute Non-disabling Cerebrovascular Events; CHARISMA, Clopidogrel for High Atherothrombotic Risk and Ischemic Stabilization, Management and Avoidance; CLAIR, Clopidogrel plus aspirin versus aspirin alone for reducing embolisation in patients with acute symptomatic cerebral or carotid artery stenosis; CNS, Central Nervous System; COMPRESS, COMbination of Clopidogrel and Aspirin for Prevention of Early REcurrence in Acute Atherothrombotic Stroke; CPSP.com, Cilostazol Stroke Prevention Study for Antiplatelet Combination; DAPT, Dual Antiplatelet Therapy; ECLIPse, Effect of Cilostazol in the Acute Lacunar Infarction Based on Pulsatility Index of Transcranial Doppler; ESPRIT, European/Australasian Stroke Prevention in Reversible Ischaemia Trial; ESPS, European Stroke Prevention Study; FASTER, Fast Assessment of Stroke and Transient Ischemic Attack to Prevent Early Recurrence; LD, Loading Dose; MATCH, Aspirin and clopidogrel compared with clopidogrel alone after recent ischaemic stroke or transient ischaemic attack in high-risk patients; MI, Myocardial Infarction; MRI, Magnetic Resonance Imaging; NA, Not Available; OD, Once Daily; POINT, Platelet-Oriented Inhibition in New TIA and Minor Ischemic Stroke; PRINCE, Platelet Reactivity in Acute Non-disabling Cerebrovascular Events; PROFESS, Prevention Regimen For Effectively Avoiding Second Strokes; SAPT, Single Antiplatelet Therapy; SPS3, Secondary Prevention of Small Subcortical Strokes; TIA, Transient Ischemic Attack; TOSS II, Trial of Cilostazol in Symptomatic Intracranial Arterial Stenosis II; VKA, Vitamin K Antagonist.

Table 6.3 Ongoing randomized trials concerning dual antiplatelet therapy for ischemic stroke treatment and prevention.

Trial (NCT)	Blinding	Country	Expected size	Population	Intervention arm	Control arm	Therapy onset	Primary endpoint	Expected follow-up	Anticipated completion date
ADANCE (NCT01924325)	Double	China	10,000	TIA or disabling non-cardioembolic ischemic stroke	Apixaban 2.5 mg twice daily or apixaban 5 mg twice daily for 21 days, then clopidogrel (75 mg OD)	ASA (75 mg OD) + clopidogrel (75 mg OD) for 21 days, then clopidogrel (75 mg OD)	≤24 h	Recurrent stroke (ischemic or hemorrhage)	90 days	July 2016
ATAMIS (NCT02869009)	Open-label	China	3,000	Mild-to-moderate ischemic stroke	ASA (100 mg OD) + clopidogrel (LD 300 mg, 75 mg OD) for 14 days, then ASA (100 mg OD) or clopidogrel (75 mg OD)	ASA (100–300 mg OD) for 14 days, then ASA (100 mg OD)	≤48 h	Early neurologic deterioration assessed as change of NIHSS	90 days	April 2020
DORIC (NCT02983214)	Open-label	Greece	1885	Patients aged ≥50 years with type 2 diabetes mellitus and symptomatic peripheral arterial disease	Clopidogrel (75 mg OD) + cilostazol (100 mg twice daily)	Clopidogrel (75 mg OD)	NA	Composite of acute ischemic stroke, TIA, MI, or death from vascular causes; BARC bleeding events	12 months	October 2019

ADANCE, Apixaban Versus Dual-antiplatelet Therapy (Clopidogrel and Aspirin) in Acute Non-disabling Cerebrovascular Events; ASA, Acetylsalicylic Acid; ATAMIS, Antiplatelet Therapy in Acute Mild-Moderate Ischemic Stroke; BARC, Bleeding Academic Research Consortium; DORIC, Diabetic Artery Obstruction: is it Possible to Reduce Ischemic Events With Cilostazol?; LD, Loading Dose; MI, Myocardial Infarction; NA, Not Applicable; NCT, National Clinical Trial; NIHSS, National Institutes of Health Stroke Scale; OD, Once Daily; TIA, Transient Ischemic Attack.

Future perspective

Among the ongoing investigations (Table 6.3), the ADANCE (Apixaban Versus Dual-antiplatelet Therapy [Clopidogrel and Aspirin] in Acute Non-disabling Cerebrovascular Events) trial is a Chinese analysis on the potential role of a direct oral anticoagulant (apixaban) for the treatment of high-risk patients with acute non-cardioembolic cerebrovascular events. Moreover, following previous RCTs in patients with acute minor stroke or TIA, the ATAMIS (Antiplatelet Therapy in Acute Mild-Moderate Ischemic Stroke) trial will provide some answers on the role of DAPT with ASA and clopidogrel among patients with a mild-to-moderate acute ischemic stroke. Finally, the DORIC (Diabetic Artery Obstruction: is it Possible to Reduce Ischemic Events With Cilostazol?) trial will explore the value of adding cilostazol to clopidogrel to prevent further ischemic events in diabetic patients with peripheral arterial disease.

Conclusion

Although DAPT has no role in the primary prevention of stroke, it plays a key role in treatment and secondary prevention strategies of ischemic non-cardioembolic cerebrovascular disease. As with any antithrombotic treatment, bleeding risk may increase and it should be carefully considered when tailoring antithrombotic therapy over the individual patient. Several aspects, particularly concerning the optimal antiplatelet regimen and its duration, remain a matter of debate and further efforts are warranted to shed some additional light on the field.

References

1. Kapil N, Datta YH, Alakbarova N, et al. Antiplatelet and anticoagulant therapies for prevention of ischemic stroke. *Clin Appl Thromb Hemost.* 2017;23(4):301–318. https://doi.org/10.1177/1076029616660762.
2. Coull AJ, Lovett JK, Rothwell PM. Population based study of early risk of stroke after transient ischaemic attack or minor stroke: implications for public education and organisation of services. *Br Med J.* 2004;328(7435):326. https://doi.org/10.1136/bmj.37991.635266.44.
3. Lavallée PC, Labreuche J, Faille D, et al. Circulating markers of endothelial dysfunction and platelet activation in patients with severe symptomatic cerebral small vessel disease. *Cerebrovasc Dis.* 2013;36(2):131–138. https://doi.org/10.1159/000353671.
4. Adams H, Bendixen B, Kappelle L, et al. Classification of subtype of acute ischemic stroke. *Stroke.* 1993;23(1):35–41. https://doi.org/10.1161/01.STR.24.1.35.
5. Hart RG, Diener H-C, Coutts SB, et al. Embolic strokes of undetermined source: the case for a new clinical construct. *Lancet Neurol.* 2014;13(4):429–438. https://doi.org/10.1016/S1474-4422(13)70310-7.

6. Powers WJ, Rabinstein AA, Ackerson T, et al. 2018 guidelines for the early management of patients with acute ischemic stroke: a guideline for healthcare professionals from the American Heart Association/American Stroke Association. 2018;49. https://doi.org/10.1161/STR.0000000000000158.

7. Wein T, Lindsay MP, Côté R, et al. Canadian stroke best practice recommendations: secondary prevention of stroke, sixth edition practice guidelines, update 2017. *Int J Stroke*. 2018;13(4):420–443. https://doi.org/10.1177/1747493017743062.

8. Wang Y, Liu M, Pu C. 2014 Chinese guidelines for secondary prevention of ischemic stroke and transient ischemic attack: compiled by the Chinese Society of Neurology, Cerebrovascular Disease Group. *Int J Stroke*. 2017;12(3):302–320. https://doi.org/10.1177/1747493017694391.

9. Piepoli MF, Hoes AW, Agewall S, et al. 2016 European guidelines on cardiovascular disease prevention in clinical practice. *Eur Heart J*. 2016;37(29):2315–2381. https://doi.org/10.1093/eurheartj/ehw106.

10. Raju N, Sobieraj-Teague M, Hirsh J, O'Donnell M, Eikelboom J. Effect of aspirin on mortality in the primary prevention of cardiovascular disease. *Am J Med*. 2011;124(7):621–629. https://doi.org/10.1016/j.amjmed.2011.01.018.

11. Antithrombotic Trialists' (ATT) Collaboration1, Baigent C, Blackwell L, et al. Aspirin in the primary and secondary prevention of vascular disease: collaborative meta-analysis of individual participant data from randomised trials. *Lancet*. 2009;373(9678):1849–1860. https://doi.org/10.1016/S0140-6736(09)60503-1.

12. Bartolucci AA, Tendera M, Howard G. Meta-analysis of multiple primary prevention trials of cardiovascular events using aspirin. *Am J Cardiol*. 2011;107(12):1796–1801. https://doi.org/10.1016/j.amjcard.2011.02.325.

13. Guirguis-Blake JM, Evans CV, Senger CA, O'Connor EA, Whitlock EP. Aspirin for the primary prevention of cardiovascular events: a systematic evidence review for the U.S. Preventive Services Task Force. *Ann Intern Med*. 2016;164(12):804. https://doi.org/10.7326/M15-2113.

14. Sloane KL, Camargo EC. Antithrombotic management of ischemic stroke. *Curr Treat Options Cardiovasc Med*. 2019;21(11):78. https://doi.org/10.1007/s11936-019-0778-4.

15. Chimowitz MI, Lynn MJ, Howlett-Smith H, et al. Comparison of warfarin and aspirin for symptomatic intracranial arterial stenosis. *N Engl J Med*. 2005;352(13):1305–1316. https://doi.org/10.1056/NEJMoa043033.

16. Del Brutto VJ, Chaturvedi S, Diener HC, Romano JG, Sacco RL. Antithrombotic therapy to prevent recurrent strokes in ischemic cerebrovascular disease: JACC scientific expert panel. *J Am Coll Cardiol*. 2019;74(6):786–803. https://doi.org/10.1016/j.jacc.2019.06.039.

17. Cadroy Y, Bossavy J-P, Thalamas C, Sagnard L, Sakariassen K, Boneu B. Early potent antithrombotic effect with combined aspirin and a loading dose of clopidogrel on experimental arterial thrombogenesis in humans. *Circulation*. 2000;101(24):2823–2828. https://doi.org/10.1161/01.CIR.101.24.2823.

18. Prasad K, Siemieniuk R, Hao Q, et al. Dual antiplatelet therapy with aspirin and clopidogrel for acute high risk transient ischaemic attack and minor ischaemic stroke: a clinical practice guideline. *Br Med J*. 2018;363. https://doi.org/10.1136/bmj.k5130.

19. Diener HC, Cunha L, Forbes C, Sivenius J, Smets P, Lowenthal A. European Stroke Prevention Study 2. Dipyridamole and acetylsalicylic acid in the secondary prevention of stroke. *J Neurol Sci*. 1996;143(1–2):1–13. https://doi.org/10.1016/S0022-510X(96)00308-5.

20. Halkes P, van Gijin J, Kappelle LJ, Koudstaal PJ, Algra A. Aspirin plus dipyridamole versus aspirin alone after cerebral ischaemia of arterial origin (ESPRIT): randomised controlled trial. *Lancet*. 2006;367(9523):1665−1673. https://doi.org/10.1016/S0140-6736(06)68734-5.

21. Sacco RL, Diener H-C, Yusuf S, et al. Aspirin and extended-release dipyridamole versus clopidogrel for recurrent stroke. *N Engl J Med*. 2008;359(12):1238−1251. https://doi.org/10.1056/NEJMoa0805002.

22. Shinohara Y, Katayama Y, Uchiyama S, et al. Cilostazol for prevention of secondary stroke (CSPS 2): an aspirin-controlled, double-blind, randomised non-inferiority trial. *Lancet Neurol*. 2010;9(10):959−968. https://doi.org/10.1016/S1474-4422(10)70198-8.

23. Toyoda K, Uchiyama S, Yamaguchi T, et al. Dual antiplatelet therapy using cilostazol for secondary prevention in patients with high-risk ischaemic stroke in Japan: a multicentre, open-label, randomised controlled trial. *Lancet Neurol*. 2019;18(6):539−548. https://doi.org/10.1016/S1474-4422(19)30148-6.

24. James SK, Storey RF, Khurmi NS, et al. Ticagrelor versus clopidogrel in patients with acute coronary syndromes and a history of stroke or transient ischemic attack. *Circulation*. 2012;125(23):2914−2921. https://doi.org/10.1161/CIRCULATIONAHA.111.082727.

25. Johnston SC, Amarenco P, Albers GW, et al. Ticagrelor versus aspirin in acute stroke or transient ischemic attack. *N Engl J Med*. 2016;375(1):35−43. https://doi.org/10.1056/NEJMoa1603060.

26. Wiviott SD, Braunwald E, McCabe CH, et al. Prasugrel versus clopidogrel in patients with acute coronary syndromes. *N Engl J Med*. 2007;357(20):2001−2015. https://doi.org/10.1056/NEJMoa0706482.

27. Chimowitz MI, Lynn MJ, Derdeyn CP, et al. Stenting versus aggressive medical therapy for intracranial arterial stenosis. *N Engl J Med*. 2011;365(11):993−1003. https://doi.org/10.1056/NEJMoa1105335.

28. Markus HS, Droste DW, Kaps M, et al. Dual antiplatelet therapy with clopidogrel and aspirin in symptomatic carotid stenosis evaluated using doppler embolic signal detection: the Clopidogrel and Aspirin for Reduction of Emboli in Symptomatic Carotid Stenosis (CARESS) trial. *Circulation*. 2005;111(17):2233−2240. https://doi.org/10.1161/01.CIR.0000163561.90680.1C.

29. Bhatt DL, Fox A, Hacke W, et al. Clopidogrel and aspirin versus aspirin alone for the prevention of atherothrombotic events. *N Engl J Med*. 2006;354(16):1706−1717. https://doi.org/10.1056/NEJMoa060989.

30. Kennedy J, Hill MD, Ryckborst KJ, Eliasziw M, Demchuk AM, Buchan AM. Fast assessment of stroke and transient ischaemic attack to prevent early recurrence (FASTER): a randomised controlled pilot trial. *Lancet Neurol*. 2007;6(11):961−969. https://doi.org/10.1016/S1474-4422(07)70250-8.

31. The investigators SPS3. Effects of clopidogrel added to aspirin in patients with recent lacunar stroke. *N Engl J Med*. 2012;367(9):817−825. https://doi.org/10.1056/NEJMoa1204133.

32. Wang Y, Wang Y, Zhao X, et al. Clopidogrel with aspirin in acute minor stroke or transient ischemic attack. *N Engl J Med*. 2013;369(1):11−19. https://doi.org/10.1056/NEJMoa1215340.

33. Johnston SC, Easton JD, Farrant M, et al. Clopidogrel and aspirin in acute ischemic stroke and high-risk TIA. *N Engl J Med*. 2018;379(3):215−225. https://doi.org/10.1056/NEJMoa1800410.

34. Pan Y, Elm JJ, Li H, et al. Outcomes associated with clopidogrel-aspirin use in minor stroke or transient ischemic attack: a pooled analysis of clopidogrel in high-risk patients with acute non-disabling cerebrovascular events (CHANCE) and platelet-oriented inhibition in new TIA and. *JAMA Neurol.* 2019;100070(119):1−8. https://doi.org/10.1001/jamaneurol.2019.2531.

35. Kheiri B, Osman M, Abdalla A, et al. Clopidogrel and aspirin after ischemic stroke or transient ischemic attack: an updated systematic review and meta-analysis of randomized clinical trials. *J Thromb Thrombolysis.* 2019;47(2):233−247. https://doi.org/10.1007/s11239-018-1786-z.

36. Ye MB, Chen YL, Wang Q, An J, Ye F, Jing P. Aspirin plus clopidogrel versus aspirin mono-therapy for ischemic stroke: a meta-analysis. *Scand Cardiovasc J.* 2019;53(4):169−175. https://doi.org/10.1080/14017431.2019.1620962.

37. Patti G, Sticchi A, Bisignani A, et al. Meta-regression to identify patients deriving the greatest benefit from dual antiplatelet therapy after stroke or transient ischemic attack without thrombolytic or thrombectomy treatment. *Am J Cardiol.* 2019;124(4):627−635. https://doi.org/10.1016/j.amjcard.2019.05.013.

38. Wang Y, Chen W, Lin Y, et al. Ticagrelor plus aspirin versus clopidogrel plus aspirin for platelet reactivity in patients with minor stroke or transient ischaemic attack: open label, blinded endpoint, randomised controlled phase II trial. *Br Med J.* 2019:l2211. https://doi.org/10.1136/bmj.l2211.

39. Johnston SC, Amarenco P, Denison H, et al. Ticagrelor and Aspirin or Aspirin Alone in Acute Ischemic Stroke or TIA. *N Engl J Med.* 2020;383:207−217. https://doi.org/10.1056/NEJMoa1916870.

40. Bath PM, Woodhouse LJ, Appleton JP, et al. Antiplatelet therapy with aspirin, clopidogrel, and dipyridamole versus clopidogrel alone or aspirin and dipyridamole in patients with acute cerebral ischaemia (TARDIS): a randomised, open-label, phase 3 superiority trial. *Lancet.* 2018;391(10123):850−859. https://doi.org/10.1016/S0140-6736(17)32849-0.

41. Eikelboom JW, Connolly SJ, Bosch J, et al. Rivaroxaban with or without aspirin in stable cardiovascular disease. *N Engl J Med.* 2017;377(14):1319−1330. https://doi.org/10.1056/NEJMoa1709118.

42. Matias-Guiu J, Davalos A, Picò M, et al. Low-dose acetylsalicylic acid (ASA) plus dipyridamole versus dipyridamole alone in the prevention of stroke in patients with reversible ischemic attacks. *Acta Neurologica Scandinavica.* 1987;76(6):413−421. https://doi.org/10.1111/j.1600-0404.1987.tb03596.x.

43. The ESPS Group. The European Stroke Prevention Study (ESPS). *The Lancet.* 1987;330(8572):1351−1354. https://doi.org/10.1016/S0140-6736(87)91254-2.

44. Diener HC, Bogousslavsky J, Brass LM, et al. Aspirin and clopidogrel compared with clopidogrel alone after recent ischaemic stroke or transient ischaemic attack in high-risk patients (MATCH): randomised, double-blind, placebo-controlled trial. *The Lancet.* 2004;364(9431):331−337. https://doi.org/10.1016/S0140-6736(04)16721-4.

45. Chairangsarit P, Sithinamsuwan P, Niyasom S. Comparison between aspirin combined with dipyridamole versus aspirin alone within 48 hours after ischemic stroke event for prevention of recurrent stroke and improvement of neurological function: a preliminary study. *J Med Assoc Thai.* 2005;88 Suppl(3):S148−S154.

46. Active Investigators. Effect of Clopidogrel Added to Aspirin in Patients with Atrial Fibrillation. *N Engl J Med.* 2009;360:2066−2078. https://doi.org/10.1056/NEJMoa0901301.

47. Wong KS, Chen C, Fu J, et al. Clopidogrel plus aspirin versus aspirin alone for reducing embolisation in patients with acute symptomatic cerebral or carotid artery stenosis (CLAIR study): a randomised, open-label, blinded-endpoint trial. *Lancet Neurol.* 2010;9(5):489−497. https://doi.org/10.1016/S1474-4422(10)70060-0.

48. Kwon SU, Hong KS, Kang DW, et al. Efficacy and Safety of Combination Antiplatelet Therapies in Patients With Symptomatic Intracranial Atherosclerotic Stenosis. *Stroke.* 2011;42:2883−2890. https://doi.org/10.1161/STROKEAHA.110.609370.

49. Nakamura T, Tsuruta S, Uchiyama S. Cilostazol combined with aspirin prevents early neurological deterioration in patients with acute ischemic stroke: a pilot study. *J Neurol Sci.* 2012;313(1−2):22−26. https://doi.org/10.1016/j.jns.2011.09.038. Epub 2011 Oct 19.

50. Han SW, Lee SS, Kim SH, et al. Effect of cilostazol in acute lacunar infarction based on pulsatility index of transcranial Doppler (ECLIPse): a multicenter, randomized, double-blind, placebo-controlled trial. *Eur Neurol.* 2012;69(1):33−40. https://doi.org/10.1159/000338247.

51. Yi X, Lin J, Wang C. A comparative study of dual versus monoantiplatelet therapy in patients with acute large-artery atherosclerosis stroke. *J Stroke Cerebrovasc Dis.* 2014; 23(7):1975−1981. https://doi.org/10.1016/j.jstrokecerebrovasdis.2014.01.022.

52. He F, Xia C, Zhang JH, et al. Clopidogrel plus aspirin versus aspirin alone for preventing early neurological deterioration in patients with acute ischemic stroke. *J Clin Neurosci.* 2015;22(1):83−86. https://doi.org/10.1016/j.jocn.2014.05.038.

53. Hong KS, Lee SH, Kim EG, et al. Recurrent Ischemic Lesions After Acute Atherothrombotic Stroke: Clopidogrel Plus Aspirin Versus Aspirin Alone. *Stroke.* 2016;47(9): 2323−2330. https://doi.org/10.1161/STROKEAHA.115.012293.

54. Aoki J, Iguchi Y, Urabe T, et al. Acute Aspirin Plus Cilostazol Dual Therapy for Non-cardioembolic Stroke Patients Within 48 Hours of Symptom Onset. *J Am Heart Assoc.* 2019;8(15):e012652. https://doi.org/10.1161/JAHA.119.012652.

Competing risks in the duration of dual antiplatelet therapy—the case for shorter treatment

Francesco Costa, MD, PhD [1], **Marco Valgimigli, MD, PhD** [2]

[1]*Department of Clinical and Experimental Medicine, Policlinic "G. Martino", University of Messina, Messina, Italy;* [2]*Swiss Cardiovascular Center Bern, Bern University Hospital, Bern, Switzerland*

Introduction

Pivotal clinical trials established dual antiplatelet therapy (DAPT) as the standard of care after percutaneous coronary intervention (PCI) and acute coronary syndrome (ACS).[1–3] International guidelines have since established 12 months as the recommended DAPT duration.[4] Reducing the duration of DAPT has been prompted by multiple factors. First, the introduction of improved drug-eluting stent (DES) design, implementing thinner struts and more biocompatible polymers, has reduced the risk of acute, subacute, late, and very late stent thrombosis and most likely the need for longer term DAPT to prevent stent-related recurrences.[5] At the same time, the increase in bleeding complications observed in the past two decades and a better understanding of their clinical consequences have raised the level of attention to these complications, with a call for safer treatment strategies.[6]

In the past decade, an exceptionally large number of studies have focused on identifying the optimal duration of DAPT, with more than 14 randomized clinical trials specifically focusing on reducing DAPT duration from the conventional 12 months after PCI to 6 or even 3 months.[7–19]

In this chapter, we will focus on the competing risks of ischemia and bleeding, as well as their clinical impact on patients' prognosis. We will also discuss the evidence in support of a shorter DAPT duration after PCI and the elements that can inform decision-making for DAPT duration individualization.

Competing risks of ischemia and bleeding: why bleeding is important and what is its clinical impact?

A residual risk of recurrent ischemic events supports the use of long-term antithrombotic therapy in patients with coronary artery disease.[20] DAPT has demonstrated the ability to reduce the risk of ischemic recurrences consistently even after the first year of treatment.[21,22] Unfortunately, the time exposure is linearly related to the excess of bleeding risk, so the longer the treatment with DAPT, the higher the probability of spontaneous bleeding.[23] Importantly, not only ischemic events can be linked to a worse outcome but also bleeding events carry a significant prognostic impact depending on their location and severity. In a substudy of the Thrombin Receptor Antagonist for Clinical Event Reduction in Acute Coronary Syndrome (TRACER) trial, the prognostic impact of out-of-hospital bleeding occurring late after hospital discharge for an ACS has been explored and compared to the impact of recurrent myocardial infarctions (MIs).[24] The impact of bleeding events on mortality was strictly related to their site and severity: nuisance bleeding that did not require medical attention was uneventful and was not associated with an increased risk of mortality; on the contrary, clinically overt, nonmajor bleeding requiring active medical evaluation carried a significant association with mortality, which was, however, smaller than the relative prognostic impact of an MI (for example, the relative risk of MI vs. bleeding type 3a [Bleeding Academic Research Consortium {BARC}] was 2.23; 95% confidence interval [CI], 1.36–3.64; $P = .001$).[24] More severe bleeding had a higher and significant impact on mortality, which was slightly lower or similar to the impact of a spontaneous MI. Intracranial or intraocular bleeding was in contrast linked to a very high risk of mortality, which was roughly five times higher than that of spontaneous MI.[24]

In another recent analysis from the PARIS registry, Sorrentino et al.[25] explored the relative impact of mortality after either ischemic or bleeding events according to the baseline bleeding risk status. Similar to what has been observed in prior studies, bleeding and ischemic events showed a similar relative prognostic impact toward mortality, and this relationship was independent of the baseline bleeding risk status of the patient.[25] The association between bleeding and mortality is complex and not completely understood. Bleeding events may be directly fatal if occurring in a vital organ, or they could indirectly lead to death following transfusions or abrupt interruption of the antithrombotic regimen or other life-saving medications. These events may also unmask a higher risk for death as a consequence of comorbidities. In addition, apart from its prognostic impact, bleeding drives poorer drug adherence and negatively affects quality of life.[26–28]

Elements associated with a high bleeding risk

In a survey of the European Association of Percutaneous Cardiovascular Intervention, physicians identified several elements believed to identify patients at high bleeding risk (HBR), prompting decisions for shorter DAPT.[29] These were older age, a history of prior bleeding, reduced renal function, anemia, and a high HAS-BLED score.[29] Older age is probably the most recognized factor of HBR after PCI, and its inclusion is ubiquitous across bleeding risk scales or among inclusion criteria in HBR-PCI trials.[30–32] Elderly patients have a higher risk of bleeding and this is further increased when prolonged treatment with DAPT is instituted. In the Prolonging Dual-Antiplatelet Treatment After Grading Stent-Induced Intimal Hyperplasia (PRODIGY) study, elderly patients treated with DAPT for 24 rather than 6 months, had an excess of BARC 2, 3, and 5, with a higher absolute bleeding risk increase than that in younger patients.[33] In addition, older patients frequently have comorbidities such as impaired renal function, anemia, and polypharmacotherapy, which jointly increase their risk of bleeding.

A history of spontaneous bleeding or blood transfusions is another important predictor of future bleeding events in clinical practice. In an analysis from the Bern PCI registry, a prior history of gastrointestinal bleeding was the single strongest predictor of recurrent bleeding events, accounting for more than threefold higher risk compared with those without prior events.[34] Anemia and impaired renal function often coexist and are both independently associated with higher risks of bleeding. Anemia could result from a current nonclinically overt bleeding, especially from the gastrointestinal tract, or from chronic comorbidities reducing bone marrow activity. Similarly, impaired renal function could cause a higher bleeding risk directly, by alteration of platelet function and coagulation cascade,[35] or indirectly, through unpredictable pharmacokinetics of antithrombotic agents and higher risk of accumulation. The bleeding risk is also increased among patients with atrial fibrillation or other conditions requiring therapy with oral anticoagulants (OACs) in conjunction with antiplatelet therapy following PCI.[36] The association of antiplatelets and OACs increases bleeding risk by threefold to fivefold.[37,38] Implementation of direct oral anticoagulants (DOACs) while withdrawing aspirin early after PCI limits this risk,[38] which however remains high, requiring a closer follow-up in this group of patients.[36]

An extensive list of clinical and biochemical HBR criteria has been proposed by a consensus of experts from the Academic Research Consortium,[39] and the list was, at least in part, validated by two independent datasets.[40,41] These elements are grouped in major and minor criteria of HBR, and accordingly, HBR status is defined if at least one major criterion or two minor criteria are present (Table 7.1).

Strategies to reduce bleeding: the case for a short dual antiplatelet therapy duration

Reducing exposure to DAPT has unequivocally demonstrated to reduce the risk of major bleeding events.[23] This relation has been studied to date in 14 randomized controlled trials (RCTs) and multiple meta-analyses (Table 7.2),[9,13,16,17,42] providing an exceptional source of evidence-based practice and informing international guidelines.[43–45] In general, these studies compared a shorter treatment duration (e.g., to 3 or 6 months) with a standard treatment (e.g., 12 months) or longer treatment (e.g., 18 or 24 months) in patients who have undergone PCI to test the hypothesis that a shorter DAPT regimen was noninferior to the standard of care in

Table 7.1 High bleeding risk criteria according to the HBR-ARC document.[8]

Major criteria
Anticipated use of long-term oral anticoagulation
Severe or end-stage CKD (eGFR <30 mL/min)
Hemoglobin <11 g/dL
Spontaneous bleeding requiring hospitalization or transfusion in the past 6 months or at any time, if recurrent
Moderate or severe baseline thrombocytopenia (platelet <100 × 10⁹/L)
Chronic bleeding diathesis
Liver cirrhosis with portal hypertension
Active malignancy
Previous spontaneous ICH, traumatic ICH within the past 12 months, presence of arterial venous malformation, moderate or severe ischemic stroke within the past 6 months
Nondeferrable major surgery on DAPT
Recent major surgery or major trauma within 30 days before PCI

Minor criteria
Age ≥75 y
Moderate CKD (eGFR 30–59 mL/min)
Hemoglobin 11–12.9 g/dL for men and 11–11.9 g/dL for women
Spontaneous bleeding requiring hospitalization or transfusion in the past 12 months not meeting the major criterion
Long-term use of oral nonsteroidal anti-inflammatory drugs or steroids
Any ischemic stroke at any time not meeting the major criterion

In order to qualify as high bleeding risk, a patient should present at least one major criterion or two minor criteria.
CKD, *chronic kidney disease;* DAPT, *dual antiplatelet therapy;* eGFR, *estimated glomerular filtration rate;* HBR-ARC, *Academic Research Consortium for High Bleeding Risk;* ICH, *intracranial haemorrhage;* PCI, *percutaneous coronary intervention.*

terms of ischemic events or net adverse clinical events (i.e., ischemic and bleeding events merged in a single composite endpoint).[4,10,12,46,47]

Eight of these RCTs have tested a shorter treatment duration of DAPT of 6 months after stenting, with a standard treatment duration for 12 months (Table 7.2). All these studies shared a similar noninferiority design.

The first study published among this group was the Efficacy of Xience/Promus versus Cypher to Reduce Late Loss After Stenting (EXCELLENT) trial.[8] This included 1443 patients treated with DES randomized to 6 versus 12 months DAPT. The trial ultimately demonstrated the noninferiority of 6 versus 12 months of DAPT with respect to the primary endpoint, a composite of cardiac death, MI, or ischemia-driven target vessel revascularization. There was a nonsignificant trend toward a reduction of Thrombolysis in Myocardial Infarction (TIMI) major and minor bleeding with the shorter course of therapy (HR 0.40, 95% CI, 0.13–1.27;

Table 7.2 Randomized controlled trials testing short-duration dual antiplatelet therapy strategies after coronary stent implantation.

	N	Year	Population	Stent type	P2Y$_{12i}$ Type	DAPT duration	Study hypothesis	Primary endpoint	Event rates (short vs. long)	Primary hypothesis met
EXCELLENT	1443	2012	ACS 51%	1st gen DES 25% 2nd gen DES 75%	Clopidogrel 100%	6 versus 12	Noninferiority of 6 mo. DAPT	Cardiac death, MI, TVR	4.8% versus 4.3%	Yes
PRODIGY	1970	2012	ACS 75%	BMS 25% 1st gen DES 25% 2nd gen DES 50%	Clopidogrel 100%	6 versus 24	Superiority of 24 mo. DAPT	All-cause death, MI, CVA	10% versus 10.1%	No
RESET	2117	2012	ACS 59%	1st gen DES 21% 2nd gen DES 85%	Clopidogrel 100%	3 versus 12	Noninferiority of 3mo. DAPT	Cardiac death, MI, ST, TVR, major bleeding	4.7% versus 4.7%	Yes
OPTIMIZE	3119	2013	ACS 35%	2nd gen DES 100%	Clopidogrel 100%	3 versus 12	Noninferiority of 3 mo. DAPT	All-cause death, MI, stroke, major bleeding	6% versus 5.8%	Yes
SECURITY	1399	2014	28% ACS	2nd gen DES 100%	Clopidogrel 98.7% Prasugrel 0.2% Ticagrelor 0.4%	6 versus 12	Noninferiority of 6 mo. DAPT	Cardiac death, MI, ST, or stroke	4.5% versus 3.7%	Yes[a]
ISAR SAFE	4000	2015	ACS 40%	1st gen DES 10% 2nd gen DES 89%	Clopidogrel 100%	6 versus 12	Noninferiority of 6 mo. DAPT	Death, MI, ST, stroke, major bleeding	1.5% versus 1.6%	Yes[a]
ITALIC	1822	2015	ACS 24%	2nd gen DES 100%	Clopidogrel 98.6% Prasugrel 1.7% Ticagrelor 0.05%	6 versus 24	Noninferiority of 6 mo. DAPT	Death, MI, TVR, stroke, major bleeding	1.6% versus 1.5%	Yes[a]

Continued

Table 7.2 Randomized controlled trials testing short-duration dual antiplatelet therapy strategies after coronary stent implantation.—cont'd

	N	Year	Population	Stent type	P2Y$_{12i}$ Type	DAPT duration	Study hypothesis	Primary endpoint	Event rates (short vs. long)	Primary hypothesis met
I LOVE IT 2	1829	2016	ACS 64%	2nd gen DES 100%	Clopidogrel 100%	6 versus 12	Noninferiority of 6 mo. DAPT	Cardiac death, TVR, MI	7.5% versus 6.3%	Yes
IVUS XPL	1400	2016	ACS 49%	2nd gen DES 100%	Clopidogrel 100%	6 versus 12	Comparison of 6 versus 12 mo. DAPT	Cardiac death, MI, stroke, or major bleeding	2.2% versus 2.1%	Yes
NIPPON	3307	2016	ACS 33%	2nd gen DES 100%	Clopidogrel 97.5% Prasugrel 0.1% Ticlopidine 2.3%	6 versus 18	Noninferiority of 6 mo. DAPT	Death, MI, CVA, major bleeding	2.1% versus 1.5%	Yes
REDUCE	1496	2017	ACS 100%	2nd gen DES 100%	Clopidogrel 40.8% Prasugrel 10.4% Ticagrelor 48.9%	3 versus 12	Noninferiority of 3 mo. DAPT	All-cause death, MI, ST, stroke, TVR, or bleeding	8.3% versus 8.5%,	Yes
DAPT-STEMI	861	2017	ACS (STEMI) 100%	2nd gen DES 100%	Clopidogrel 42.0% Prasugrel 29.5% Ticagrelor 28.5%	6 versus 12	Noninferiority of 6 mo. DAPT	All-cause mortality, MI, revascularization, stroke, and TIMI major bleeding	4.8% versus 6.6%	Yes
OPTIMA-C	1368	2018	ACS 50%	2nd gen DES 100%	Clopidogrel 100%	6 versus 12	Noninferiority of 6 versus 12 mo. DAPT	Cardiac death, TVR MI, ischemia-driven TVR	1.2% versus 0.6%	Yes
SMART-DATE	2712	2018	ACS 100%	2nd gen DES 100%	Clopidogrel 80.7% Prasugrel/ ticagrelor 19.3%	6 versus 12	Noninferiority of 6 mo. DAPT	All-cause mortality, MI, stroke	4.7% versus 4.2%	Yes

ACS, acute coronary syndrome; BMS, bare metal stent; CVA, cerebrovascular accident; DAPT, dual antiplatelet therapy; DES, drug-eluting stent; MI, myocardial infarction; MO, months; ST, stent thrombosis; STEMI, ST-elevation myocardial infarction; TIMI, Thrombolysis in Myocardial Infarction; TVR, target vessel revascularization.
[a] Terminated prematurely.

$P = .12$).[8] These results were largely consistent with the other clinical trials presented in later years, and meta-analysis confirmed that a shorter treatment duration for 6 months after stenting with a DES is associated to a significant reduction of major bleeding with no increase in coronary ischemic events.[48]

More recently two randomized trials tested a shorter DAPT duration of 6 months exclusively in the setting of ACS, a subset of patients with particular risk for recurrent ischemic events. The SMART-DATE randomized 2712 patients undergoing PCI for an ACS in South Korea to DAPT for 6 or 12 months.[18] All classes of ACS were well represented in the population: one-third presented with ST-elevation MI (STEMI), one-third presented with non-ST-elevation MI (NSTEMI), and one-third had unstable angina. The study found no difference in the rates of the primary endpoint between the two study arms, a composite of all-cause death, MI, or stroke, determining noninferiority of the experimental treatment. Yet, a significant excess of MI was noted in the short DAPT arm, with events accruing after 6 months (HR = 5.06 [1.46−17.5], $P = .01$).

The DAPT-STEMI trial[7] included 870 patients with STEMI treated with primary PCI and second-generation DES. According to the study design, after 6 months of uneventful treatment with DAPT, patients were randomized to continue DAPT up to 12 months or to stop the $P2Y_{12}$ inhibitor continuing only with aspirin. The primary study endpoint was a composite of death, MI, revascularization, stroke, and major bleeding at 24 months after primary PCI. Short DAPT was found to be noninferior as compared to the standard 12-month treatment duration (short DAPT 4.8% vs. long DAPT 6.6%; $P_{noninferiority} = .004$). While these studies may still be underpowered when taken singularly, a network meta-analysis combining 17 trials confirmed a similar efficacy and improved safety of a 6-month DAPT regimen among patients presenting with ACS.[49]

Six-month treatment duration was compared to >12-month treatment duration in two noninferiority studies, i.e., the Is There A LIfe for DES After Discontinuation of Clopidogrel (ITALIC)[10] and Nobori Dual Antiplatelet Therapy as Appropriate Duration (NIPPON)[12] trials (Table 7.1), as well as in one superiority study, i.e., the PRODIGY trial. In the ITALIC and NIPPON trials, patients were randomly allocated to 6 versus 24 months and 6 versus 18 months of DAPT, respectively. Due to slow enrollment, both studies were prematurely terminated reducing the ability to glean firm conclusions; yet, taking into account these limitations, both studies met the prespecified noninferiority margin.[12] The PRODIGY trial tested in a 4:2 factorial design the effect of four coronary stents and two DAPT duration strategies (6 vs. 24 months of DAPT) in 2013 all-comer patients who have undergone PCI.[14] At 24 months' follow-up, there was no difference in the primary efficacy endpoint of death, MI, or stroke between 6 and 24 months of DAPT, hence not reaching the primary hypothesis of the study of superiority of the longer treatment duration. Instead, an excess of BARC 2, 3, or 5 bleeding events was observed in patients treated in the longer DAPT duration arm.[14] In addition, three randomized studies tested an even shorter DAPT duration, lasting 3 months after DES implantation.[13,16] The Real Safety and Efficacy of 3-Month Dual Antiplatelet Therapy Following Endeavor

ZES Implantation (RESET) and Optimized Duration of Clopidogrel Therapy Following Treatment With the Zotarolimus-Eluting Stent in Real-World Clinical Practice (OPTIMIZE)[13] trials included a total of 2117 and 3119 patients, respectively, and both studies demonstrated the noninferiority of 3 versus 12 months of DAPT for net adverse clinical events (NACE, a summation of ischemic and bleeding events).[13,16] It is important to highlight that all the patients enrolled in these two trials were treated with the Endevor zotarolimus-eluting stent, which is no longer available on the market.

More recently, the Randomized Evaluation of Short-term DUal Anti Platelet Therapy in Patients With Acute Coronary Syndrome Treated With the COMBO Dual-therapy stEnt (REDUCE) trial, tested among patients with ACS treated with PCI the noninferiority of 3 versus 12 months of DAPT. The study included 1496 patients treated exclusively with a bioabsorbable polymer DES with a luminal CD34$^+$ antibody coating (presented in 2017, not yet published in peer-reviewed journal). The primary endpoint, a composite of all-cause death, MI, stent thrombosis, stroke, target vessel revascularization, or bleeding, was observed in 8.2% of patients in the short DAPT arm and 8.4% in the long DAPT arm, demonstrating noninferiority of the experimental treatment ($P_{noninferiority} < .001$). Yet, the study design was characterized by a wide noninferiority margin of 5%. Exploring other secondary ischemic endpoints, a borderline increase in stent thrombosis (1.2% vs. 0.4%; $P = .08$) and all-cause mortality (1.9% vs. 0.8%; $P = .07$) in the short DAPT arm was however observed. In fact, in a meta-analysis of 11,473 patients, after ACS presentation, 3 months of DAPT, but not 6 months of DAPT, was associated with a significant increase in MI or stent thrombosis.[48]

In conclusion, these studies taken together support a beneficial profile of a shorter term DAPT for 6 months, with a substantial equipoise for ischemic events and a reduction of major bleeding events. An even shorter term DAPT of 3 months appears feasible for selected patients with stable CAD. Short treatment with DAPT for 6 months, but not 3 months, appeared safe and effective also in the ACS setting. Yet, as patients with ACS appear also to benefit long term from prolonged treatment with DAPT,[50] individualization of treatment to optimize treatment benefits and risks seems mandatory.

Finally, a series of studies among patients considered at HBR have tested the feasibility of a very short DAPT duration for just 1 month (Table 7.3).[31,32,51,52] While these studies did not randomize patients to DAPT duration but to a stent type, no safety or efficacy considerations could be formulated to support preference of 1-month DAPT versus 3—6 or 12 months of treatment in patients at HBR. In fact, compared to other PCI trials, the rate of ischemic complications reported in HBR-PCI trials appears higher, including higher rates of stent thrombosis.[32,52] Whether this is due to the higher baseline ischemic risk of the HBR population or an excess of ischemic complications due to the very short duration of DAPT is unknown. Future trials of patients at HBR randomized to DAPT duration will shed light on this matter (Table 7.3).[53]

Table 7.3 Randomized controlled trials testing novel stent platforms among patients at HBR requiring short-term dual antiplatelet therapy.

	N	Year	Dominant criteria for HBR	Randomized DAPT	DAPT duration	Stent type	Study hypothesis	Primary endpoint	Study results
Completed									
ZEUS HBR	828	2015	Elderly (i.e., >80 yr) = 51.3% Need for OAC = 37.5%	No	1 mo.	Durable polymer zotarolimus DES (ENDEAVOR) 50% Bare metal stent 50%	Superiority of DES versus BMS	Death, MI, or TVR	22.6% versus 29% P_{super} = .03
LEADERS-FREE	2466	2015	Elderly (i.e., >75 yr) = 64.3% Need for OAC = 36.1%	No	1 mo.	Polymer-free Biolimus A9 DES (BIOFREEDOM) 50% Bare metal stent 50%	Noninferiority of DCS versus BMS	Cardiac death, MI, or ST (def/prob)	9.4% versus 12.9% P_{noninf} < .001 P_{super} = .005
SENIOR	1200	2018	Elderly (i.e., >75 yr) = 100%	No	1 mo.	Bioresorbable polymer everolimus DES (SYNERGY) 50% Bare metal stent 50%	Superiority of DES versus BMS	Death, MI, stroke, or TLR	12% versus 16% P_{super} = .02
ONYX ONE	1996	2020	Elderly (i.e., >75 yr) = 61.7% Need for OAC = 38.5%	No	1 mo.	Durable polymer zotarolimus DES (ONYX) 50% Polymer-free Biolimus A9 DES	Noninferiority of Onyx versus Biofreedom stent	Cardiac death, MI, or ST (def/prob)	17.1% versus 16.9% P_{noninf} = .01

Continued

Table 7.3 Randomized controlled trials testing novel stent platforms among patients at HBR requiring short-term dual antiplatelet therapy.—cont'd

	N	Year	Dominant criteria for HBR	Randomized DAPT	DAPT duration	Stent type	Study hypothesis	Primary endpoint	Study results
Ongoing									
						(BIOFREEDOM) 50%			
MASTER-DAPT (NCT03023020)	4300	Expected 2020	N.A.	Yes	1 versus 6–12 mo.	Bioresorbable polymer sirolimus DES (ULTIMASTER) 100%	Noninferiority of 1 mo. versus 6 –12 mo. DAPT	Death, MI, stroke. and BARC 3 or 5 bleeding	
TARGET SAFE (NCT03287167)	1720	Expected 2021	N.A.	Yes	1 versus 6 mo.	Bioresorbable polymer sirolimus DES (FIREHAWK) 100%	Noninferiority of 1 versus 6 mo. DAPT	Death, MI, CVA, and major bleeding	
COBRA REDUCE (NCT02594501)	996	Expected 2020	N.A.	Yes (randomization linked with stent type)	2 wk versus 3–6 mo.	Nano-coated BMS (COBRA PZF) 50% 2nd generation DES 50%	Noninferiority of 1 versus 6 mo. DAPT	Death, MI, stroke, def/ prob ST	

BARC, *Bleeding academic research consortium;* BMS, *bare metal stent;* CVA, *cerebrovascular accident;* DAPT, *dual antiplatelet therapy;* DCS, *drug-coated stent;* DES, *drug-eluting stent;* HBR, *high bleeding risk;* MI, *myocardial infarction;* MO, *months;* OAC, *oral anticoagulant;* ST, *stent thrombosis;* TLR, *target lesion revascularization;* TVR, *target vessel revascularization.*

Current evidence for individualization of duration of dual antiplatelet therapy

Results from multiple RCTs suggest that shortening treatment with DAPT is associated with a reduction of minor and major bleeding, but long-term antithrombotic therapy provides additional benefit in terms of reduction of coronary and noncoronary ischemic events in selected patients.[30] Susceptibility to spontaneous bleeding during DAPT varies among patients based on individual characteristics. Hence, selecting patients at HBR upfront after PCI may reduce the risk of complications in this specific subgroup, whereas in those who are not deemed at HBR a standard or even prolonged treatment with DAPT may be considered to provide additional ischemic protection.

Bleeding risk scores are clinical prediction tools that estimate the risk of bleeding taking into account multiple predictors at the same time. These tools have long been used to inform treatment decisions for OAC use in patients with atrial fibrillation, and recently several dedicated scores have also been proposed to do the same for DAPT duration decisions.[30,44] The first attempt to implement a bleeding risk score to inform DAPT duration decision-making was a subgroup analysis from the PRODIGY trial.[54] Patients were stratified according to their bleeding risk at baseline as HBR (Can Rapid risk stratification of Unstable angina patients Suppress Adverse outcomes with Early implementation of the ACC/AHA Guidelines [CRUSADE] score >40) or non-HBR (CRUSADE score ≤40). Patients deemed at HBR had a significant increase in major bleeding and red blood cell transfusion when treated with 24- versus 6-month DAPT (9.7% vs. 3.7%; Absolute risk difference (ARD) 6%; 95% CI, 0.4%−12.3%; $P = .04$; number needed to treat to harm = 17), whereas those *not* deemed at HBR were not exposed to a significant excess of bleeding even with long-term DAPT (2.4% vs. 1.6%; ARD 0.8%; 95% CI, −0.6%−2.2%; $P = .25$) ($P_{int} = .05$).[54] However, the CRUSADE score had many limitations for long-term bleeding risk stratification in patients who have undergone PCI. First, the CRUSADE score was originally generated to predict in-hospital bleeding events, whereas most bleeding during DAPT are occurring late after discharge. Indeed, many of the bleeding risk predictors of the CRUSADE score (e.g., blood pressure at presentation, acute heart failure at presentation) do not appear appropriate for long-term bleeding prediction.

To overcome these limitations, the Predicting Bleeding Complication in Patients Undergoing Stent Implantation and Subsequent Dual Antiplatelet Therapy (PRECISE-DAPT) score was generated from a pooled dataset of eight RCTs including a total of 14,963 patients treated with PCI and subsequent DAPT (Fig. 7.1).[55] The aim of this tool was to estimate the long-term bleeding risk while on DAPT and develop a predicting tool to inform DAPT duration. After multivariable modeling, five clinical and laboratory factors emerged as independent predictors of out-of-hospital bleeding and have been included in the score (Fig. 7.1). The score showed good internal validity (Fig. 7.2) and was further validated in two independent external cohorts of 8595 from the PLATO trial and 6172 patients from the Bern PCI registry.[3,34]

PRECISE-DAPT Score

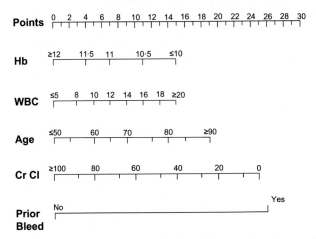

FIGURE 7.1 The predicting bleeding complication in patients undergoing stent implantation and subsequent dual Antiplatelet therapy (PRECISE-DAPT) score nomogram.

Nomogram for score calculation is based on five predictors. *CrCl*, creatinine clearance; *Hb*, hemoglobin; *WBC*, white blood cell.

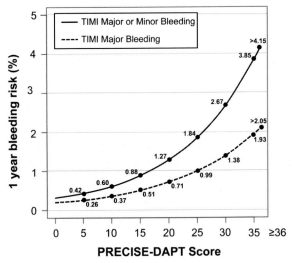

FIGURE 7.2 The predicting bleeding complication in patients undergoing stent implantation and subsequent dual antiplatelet therapy (PRECISE-DAPT) score bleeding risk prediction.

Risk curves for out-of-hospital major or minor bleeding (*solid line*) or major bleeding (*dashed line*) while on treatment with dual antiplatelet therapy. *TIMI*, Thrombolysis in Myocardial Infarction.

Among patients randomly allocated to short (3 or 6 months) or long (12 or 24 months) DAPT duration (N = 10,081), patients deemed at HBR based on a PRE-CISE DAPT score ≥25, a significant increase in TIMI major or minor bleeding was observed with long DAPT, as compared to short DAPT (ARD +2.59% [+0.82 to +4.34]; number needed to treat to harm: 38). In contrast, non-HBR patients with a PRECISE-DAPT score <25 had a significant reduction in the composite ischemic endpoint of MI, definite stent thrombosis, stroke, or target vessel revascularization (ARD −1.53% [−2.64 to −0.41]; number needed to treat: 65), but not in those at HBR (ARD +1.41% [−1.67 to +4.50]; $P_{interaction}$ = .07), without excess of bleeding complications (ARD +0.14% [−0.22 to +0.49]; $P_{interaction}$ = .007)

Irrespective of their baseline characteristics, patients with a need for long-term OAC are considered per se at HBR. These patients have been invariably excluded from randomized clinical trials for DAPT duration, which limit our ability to formulate informed decisions. Current knowledge is based on dedicated clinical trials that tested DOACs versus vitamin K antagonists.[56−59] These trials concurrently tested withdrawal of aspirin, hence early interruption of DAPT after PCI, demonstrating a significant reduction of major bleeding events. Early aspirin withdrawal (i.e., 1−4 weeks) after PCI and dual therapy with a $P2Y_{12}$ inhibitor with a DOAC is now considered the standard of care in this population.

However, there is still uncertainty regarding the optimal duration of DAPT in association to OAC after PCI, as this was not specifically explored in these pivotal trials. The Intracoronary Stenting and Antithrombotic Regimen: Testing of a Six-week Versus a Six-month Clopidogrel Treatment Regimen in Patients With Concomitant Aspirin and Oral Anticoagulant Therapy Following Drug-eluting Stenting (ISAR-TRIPLE) study randomized 614 patients requiring OACs to 6 weeks versus 6 months of DAPT after stent implantation.[60] The primary endpoint of death, MI, stent thrombosis, ischemic stroke, or TIMI major bleeding was similar in the two study arms (6 wk: 9.8% vs. 6 mo: 8.8%; HR 1.14, 95% CI, 0.68−1.91; P = .63).[60] Surprisingly, no difference in TIMI major bleeding (5.3% vs. 4.0%; HR 1.35, 95% CI, 0.64−2.84; P = .44) was observed. No definitive conclusion could be drawn given the very small sample size of this study, and more studies, focused on the duration of DAPT after stenting in patients taking OACs, are needed.

Additional stratification of risk based on bleeding in patients with AF-PCI has been recently tested in a subgroup analysis of the Randomized Evaluation of Dual Antithrombotic Therapy with Dabigatran versus Triple Therapy with Warfarin in Patients with Nonvalvular Atrial Fibrillation Undergoing Percutaneous Coronary Intervention (RE-DUAL PCI) trial and presented at the American Heart Association meeting.[61] Patients randomly assigned to double therapy with dabigatran 110 mg or 150 mg plus a $P2Y_{12}$ inhibitor or triple therapy with warfarin, aspirin, and a $P2Y_{12}$ inhibitor were stratified according to their baseline bleeding risk based on the PRECISE-DAPT score (i.e., high risk score ≥25 points). Double therapy with dabigatran 110 mg was associated with lower risk of International Society on Thrombosis and Hemostasis (ISTH) major or clinically relevant bleeding events compared to triple therapy. This was observed both in non-HBR (HR 0.42, 95%

CI, 0.31–0.57) and HBR (HR 0.70, 95% CI, 0.52–0.94) patients, with apparently higher benefit among those at non-HBR (P_{int} = .017). Dual therapy with dabigatran 150 mg versus triple therapy was associated with reduced bleeding risk in non-HBR patients (HR 0.60, 95% CI, 0.45–0.80) but not among patients at HBR (HR 0.92, 95% CI, 0.63–1.34, P_{int} = .082). While these results should be considered hypothesis generating, implementing the baseline bleeding risk to inform both duration of DAPT and OAC dosage in patients who have undergone PCI needing chronic OAC is promising and may merit further study in dedicated prospective trials.

Dual antiplatelet therapy decision-making in patients with concurrent high ischemic and bleeding risk

While there seem to be a direct relationship between bleeding risk and duration of DAPT, the same conclusion could not be drawn for ischemic events. Longer DAPT duration, which appears to reduce recurrent ischemic events in patients at higher ischemic risk, may not provide the same effect in subsets at lower ischemic risk. In fact, more intense antithrombotic therapy in patients at lower ischemic risk appeared to provide no benefit or even an increase of ischemic complications. In a subgroup analysis from the PRODIGY trial, patients with stable CAD slated for longer duration DAPT showed a fivefold increased risk of BARC 2, 3, or 5 bleeding but also a trend toward and excess of ischemic events.[62] Similarly, in a subgroup analysis from the CHARISMA trial, patients at lower ischemic risk treated with an intensified antithrombotic regimen showed a 68% increase in Global Utilization of Streptokinase and Tissue Plasminogen Activator for Occluded Coronary Arteries (GUSTO) severe bleeding and a concomitant increase of 30% for all-cause mortality and 44% of cardiovascular mortality.[63] Several explanations have been proposed to explain these findings: excessive antithrombotic therapy could be directly linked to bleeding events that could drive subsequent ischemic events by temporary or permanent discontinuation of secondary prevention medications.[26] A more speculative explanation suggested that potent or more prolonged therapy could induce atherosclerotic plaque hemorrhage and plaque rupture leading to recurrent coronary or cerebrovascular ischemic events.[64,65] A similar relationship has been observed in the Rotterdam study.[66] Among 1740 patients with carotid atherosclerosis evaluated with magnetic resonance imaging, current or past use of vitamin K antagonists or antiplatelet agents was associated with a trend toward higher incidence of carotid intraplaque hemorrhage. Specifically, there was a significant threefold increase in intraplaque hemorrhage with vitamin K antagonists (odds ratio [OR] 3.15, 95% CI, 1.23–8.05) and a 50% increase with antiplatelet agents (OR 1.50, 95% CI, 1.21–1.87).

Treatment decisions are even more difficult when both ischemic and bleeding risk are high. Evidence in this setting is scant, given the limited data available in this group of patients. In addition, exploration of multiple subgroups in clinical trials increases the risk of type 1 error, mandating caution in interpretation. In a subgroup analysis of the PRECISE-DAPT study limited to patients with ACS, longer treatment with DAPT was beneficial only among patients considered at low bleeding risk, whereas among individuals at HBR, despite the high baseline ischemic risk conveyed by ACS presentation, no benefit for ischemic events was observed. In turn, an excess of bleeding events was observed in this group of patients when treated with longer term DAPT, suggesting a benefit from shorter term DAPT irrespective of the initial ACS status in patients at HBR.[55]

Similarly, another analysis focused on the intersection of high ischemic and bleeding risk, stratifying the population at HBR or non-HBR based on the PRECISE-DAPT score and at high ischemic or low ischemic risk based on PCI complexity using a standardized definition.[67] Consistently, patients deemed non-HBR had significant ischemic benefit regardless of ischemic risk (complex PCI-ARD: −3.86%; 95% CI, −7.71 to +0.06 and noncomplex PCI-ARD: −1.14%; 95% CI, −2.26 to −0.02) In patients at HBR, there was no ischemic benefit irrespective of PCI complexity. On the other hand, prolonged DAPT increased risk of bleeding only in patients at HBR, irrespective of ischemic risk.[67]

In conclusion, clinical evidence for patients at both high bleeding and ischemic risk is still limited to subgroup analyses of randomized clinical trials. The current evidence suggests that patients at HBR, irrespective of their clinical presentation or the complexity of the coronary anatomy, do not appear to gain any additional benefit from longer DAPT and may be better targeted with a shorter treatment duration in order to avoid bleeding. Hence, bleeding risk should receive priority over ischemic risk to inform treatment decision-making for DAPT duration.

Conclusions

DAPT after PCI or ACS has an important role in reducing recurrent ischemic events but is associated with a higher rate of bleeding, which increases with the length of DAPT duration. Major bleeding events worsen prognosis and should be prevented. Shortening DAPT duration has been shown to reduce bleeding events and appears particularly attractive among patients with a higher baseline bleeding risk. In general, shorter DAPT duration seems safe and noninferior to longer DAPT duration with respect to recurrent ischemic events.

References

1. Yusuf S, Zhao F, Mehta SR, et al. Effects of clopidogrel in addition to aspirin in patients with acute coronary syndromes without ST-segment elevation. *N Engl J Med.* 2001; 345(7):494−502.
2. Wiviott SD, Braunwald E, McCabe CH, et al. Prasugrel versus clopidogrel in patients with acute coronary syndromes. *N Engl J Med.* 2007;357(20):2001−2015.

3. Wallentin L, Becker RC, Budaj A, et al. Ticagrelor versus clopidogrel in patients with acute coronary syndromes. *N Engl J Med.* 2009;361(11):1045−1057.
4. Task Force on Myocardial Revascularization of the European Society of C, the European Association for Cardio-Thoracic S, European Association for Percutaneous Cardiovascular I, et al. Guidelines on myocardial revascularization. *Eur Heart J.* 2010;31(20): 2501−2555.
5. Stone GW, Rizvi A, Newman W, et al. Everolimus-eluting versus paclitaxel-eluting stents in coronary artery disease. *N Engl J Med.* 2010;362(18):1663−1674.
6. Simonsson M, Wallentin L, Alfredsson J, et al. Temporal trends in bleeding events in acute myocardial infarction: insights from the SWEDEHEART registry. *Eur Heart J.* 2020;41(7):833−843.
7. Kedhi E, Fabris E, van der Ent M, et al. Six months versus 12 months dual antiplatelet therapy after drug-eluting stent implantation in ST-elevation myocardial infarction (DAPT-STEMI): randomised, multicentre, non-inferiority trial. *BMJ.* 2018;363:k3793.
8. Gwon H-C, Hahn J-Y, Park KW, et al. Six-month versus 12-month dual antiplatelet therapy after implantation of drug-eluting stents: the efficacy of Xience/Promus versus cypher to reduce late loss after stenting (EXCELLENT) randomized, multicenter study. *Circulation.* 2012;125(3):505−513. PubMed comprises more than 30 million citations for biomedical literature from MEDLINE, Life Science Journals, and Online Books.
9. Han Y, Xu B, Xu K, et al. Six versus 12 Months of dual antiplatelet therapy after implantation of biodegradable polymer sirolimus-eluting stent: randomized substudy of the I-LOVE-IT 2 trial. *Circ Cardiovasc Interv.* 2016;9(2):e003145.
10. Gilard M, Barragan P, Noryani AA, et al. 6- versus 24-month dual antiplatelet therapy after implantation of drug-eluting stents in patients nonresistant to aspirin: the randomized, multicenter ITALIC trial. *J Am Coll Cardiol.* 2015;65(8):777−786.
11. Hong S-J, Shin D-H, Kim J-S, et al. 6-Month versus 12-month dual-antiplatelet therapy following long everolimus-eluting stent implantation: the IVUS-XPL randomized clinical trial. *JACC Cardiovasc Interv.* 2016;9(14):1438−1446.
12. Nakamura M, Iijima R, Ako J, et al. Dual antiplatelet therapy for 6 versus 18 Months after biodegradable polymer drug-eluting stent implantation. *JACC Cardiovasc Interv.* 2017;10(12):1189−1198.
13. Feres F, Costa RA, Abizaid A, et al. Three vs twelve months of dual antiplatelet therapy after zotarolimus-eluting stents: the OPTIMIZE randomized trial. *Jama.* 2013;310(23): 2510−2522.
14. Valgimigli M, Campo G, Monti M, et al. Short- versus long-term duration of dual-antiplatelet therapy after coronary stenting: a randomized multicenter trial. *Circulation.* 2012;125(16):2015−2026.
15. De Luca G, Damen SA, Camaro C, et al. Final results of the randomised evaluation of short-term dual antiplatelet therapy in patients with acute coronary syndrome treated with a new-generation stent (REDUCE trial). *EuroIntervention.* 2019;15(11):e990−e998.
16. Kim BK, Hong MK, Shin DH, et al. A new strategy for discontinuation of dual antiplatelet therapy: the RESET Trial (REal safety and efficacy of 3-month dual antiplatelet therapy following Endeavor zotarolimus eluting stent implantation). *J Am Coll Cardiol.* 2012;60(15):1340−1348.
17. Colombo A, Chieffo A, Frasheri A, et al. Second-generation drug-eluting stent implantation followed by 6- versus 12-month dual antiplatelet therapy: the SECURITY randomized clinical trial. *J Am Coll Cardiol.* 2014;64(20):2086−2097.

18. Hahn JY, Song YB, Oh JH, et al. 6-month versus 12-month or longer dual antiplatelet therapy after percutaneous coronary intervention in patients with acute coronary syndrome (SMART-DATE): a randomised, open-label, non-inferiority trial. *Lancet*. 2018; 391(10127):1274–1284.

19. Lee BK, Kim JS, Lee OH, et al. Safety of six-month dual antiplatelet therapy after second-generation drug-eluting stent implantation: OPTIMA-C randomised clinical trial and OCT substudy. *EuroIntervention*. 2018;13(16):1923–1930.

20. Bhatt DL, Steg PG, Ohman EM, et al. International prevalence, recognition, and treatment of cardiovascular risk factors in outpatients with atherothrombosis. *Jama*. 2006; 295(2):180–189.

21. Mauri L, Kereiakes DJ, Yeh RW, et al. Twelve or 30 Months of dual antiplatelet therapy after drug-eluting stents. *N Engl J Med*. 2014;371(23):2155–2166.

22. Bonaca MP, Bhatt DL, Cohen M, et al. Long-term use of ticagrelor in patients with prior myocardial infarction. *N Engl J Med*. 2015;372(19):1791–1800.

23. Navarese EP, Andreotti F, Schulze V, et al. Optimal duration of dual antiplatelet therapy after percutaneous coronary intervention with drug eluting stents: meta-analysis of randomised controlled trials. *BMJ*. 2015;350:h1618.

24. Valgimigli M, Costa F, Lokhnygina Y, et al. Trade-off of myocardial infarction vs. bleeding types on mortality after acute coronary syndrome: lessons from the thrombin receptor antagonist for clinical event reduction in acute coronary syndrome (TRACER) randomized trial. *Eur Heart J*. 2017;38(11):804–810.

25. Sorrentino S, Sartori S, Baber U, et al. Bleeding risk, dual antiplatelet therapy cessation, and adverse events after percutaneous coronary intervention: the PARIS registry. *Circ Cardiovasc Interv*. 2020;13(4):e008226.

26. Mehran R, Baber U, Steg PG, et al. Cessation of dual antiplatelet treatment and cardiac events after percutaneous coronary intervention (PARIS): 2 year results from a prospective observational study. *Lancet*. 2013;382(9906):1714–1722.

27. Amin AP, Bachuwar A, Reid KJ, et al. Nuisance bleeding with prolonged dual antiplatelet therapy after acute myocardial infarction and its impact on health status. *J Am Coll Cardiol*. 2013;61(21):2130–2138.

28. Valgimigli M, Garcia-Garcia HM, Vrijens B, et al. Standardized classification and framework for reporting, interpreting, and analysing medication non-adherence in cardiovascular clinical trials: a consensus report from the Non-adherence Academic Research Consortium (NARC). *Eur Heart J*. 2019;40(25):2070–2085.

29. Valgimigli M, Costa F, Byrne R, Haude M, Baumbach A, Windecker S. Dual antiplatelet therapy duration after coronary stenting in clinical practice: results of an EAPCI survey. *EuroIntervention*. 2015;11(1):68–74.

30. Costa F, Windecker S, Valgimigli M. Dual antiplatelet therapy duration: reconciling the inconsistencies. *Drugs*. 2017;77(16):1733–1754.

31. Varenne O, Cook S, Sideris G, et al. Drug-eluting stents in elderly patients with coronary artery disease (SENIOR): a randomised single-blind trial. *Lancet*. 2018;391(10115): 41–50.

32. Windecker S, Latib A, Kedhi E, et al. Polymer-based or polymer-free stents in patients at high bleeding risk. *N Engl J Med*. 2020;382(13):1208–1218.

33. Piccolo R, Magnani G, Ariotti S, et al. Ischaemic and bleeding outcomes in elderly patients undergoing a prolonged versus shortened duration of dual antiplatelet therapy after percutaneous coronary intervention: insights from the PRODIGY randomised trial. *EuroIntervention*. 2017;13(1):78–86.

34. Koskinas KC, Raber L, Zanchin T, et al. Clinical impact of gastrointestinal bleeding in patients undergoing percutaneous coronary interventions. *Circ Cardiovasc Interv.* 2015; 8(5).

35. Lutz J, Menke J, Sollinger D, Schinzel H, Thurmel K. Haemostasis in chronic kidney disease. *Nephrol Dial Transplant.* 2014;29(1):29−40.

36. van Rein N, Heide-Jorgensen U, Lijfering WM, Dekkers OM, Sorensen HT, Cannegieter SC. Major bleeding rates in atrial fibrillation patients on single, dual, or triple antithrombotic therapy. *Circulation.* 2019;139(6):775−786.

37. Lamberts M, Gislason GH, Lip GY, et al. Antiplatelet therapy for stable coronary artery disease in atrial fibrillation patients taking an oral anticoagulant: a nationwide cohort study. *Circulation.* 2014;129(15):1577−1585.

38. Ando G, Costa F. Double or triple antithrombotic therapy after coronary stenting and atrial fibrillation: a systematic review and meta-analysis of randomized clinical trials. *Int J Cardiol.* 2020;302:95−102.

39. Urban P, Mehran R, Colleran R, et al. Defining high bleeding risk in patients undergoing percutaneous coronary intervention. *Circulation.* 2019;140(3):240−261.

40. Natsuaki M, Morimoto T, Shiomi H, et al. Application of the academic research consortium high bleeding risk criteria in an all-comers registry of percutaneous coronary intervention. *Circ Cardiovasc Interv.* 2019;12(11):e008307.

41. Ueki Y, Bar S, Losdat S, et al. Validation of bleeding risk criteria (ARC-HBR) in patients undergoing percutaneous coronary intervention and comparison with contemporary bleeding risk scores. *EuroIntervention.* 2020;16(5):371−379.

42. Schulz-Schupke S, Byrne RA, Ten Berg JM, et al. ISAR-SAFE: a randomized, double-blind, placebo-controlled trial of 6 vs. 12 months of clopidogrel therapy after drug-eluting stenting. *Eur Heart J.* 2015;36(20):1252−1263.

43. Marquis-Gravel G, Metha S, Valgimigli M, et al. A critical comparison of Canadian and international guidelines recommendations for antiplatelet therapy in coronary artery disease. *Can J Cardiol.* 2020;36(8):1298−1307.

44. Valgimigli M, Bueno H, Byrne RA, et al. 2017 ESC focused update on dual antiplatelet therapy in coronary artery disease developed in collaboration with EACTS: the Task Force for dual antiplatelet therapy in coronary artery disease of the European Society of Cardiology (ESC) and of the European Association for Cardio-Thoracic Surgery (EACTS). *Eur Heart J.* 2018;39(3):213−260.

45. Levine GN, Bates ER, Bittl JA, et al. 2016 ACC/AHA guideline focused update on duration of dual antiplatelet therapy in patients with coronary artery disease: a report of the American College of Cardiology/American Heart Association Task Force on clinical practice guidelines. *J Am Coll Cardiol.* 2016;68(10):1082−1115.

46. Authors/Task Force m, Windecker S, Kolh P, et al. 2014 ESC/EACTS guidelines on myocardial revascularization: the task force on myocardial revascularization of the European Society of Cardiology (ESC) and the European Association for Cardio-Thoracic Surgery (EACTS)developed with the special contribution of the European Association of Percutaneous Cardiovascular Onterventions (EAPCI). *Eur Heart J.* 2014;35(37): 2541−2619.

47. Costa F, Ariotti S, Valgimigli M, et al. Perspectives on the 2014 ESC/EACTS guidelines on myocardial revascularization : fifty years of revascularization: where are we and where are we heading? *J Cardiovasc Transl Res.* 2015;8(4):211−220.

48. Palmerini T, Della Riva D, Benedetto U, et al. Three, six, or twelve months of dual antiplatelet therapy after DES implantation in patients with or without acute coronary

syndromes: an individual patient data pairwise and network meta-analysis of six randomized trials and 11 473 patients. *Eur Heart J.* 2017;38(14):1034−1043.

49. Yin SH, Xu P, Wang B, et al. Duration of dual antiplatelet therapy after percutaneous coronary intervention with drug-eluting stent: systematic review and network meta-analysis. *BMJ.* 2019;365:l2222.

50. Udell JA, Bonaca MP, Collet JP, et al. Long-term dual antiplatelet therapy for secondary prevention of cardiovascular events in the subgroup of patients with previous myocardial infarction: a collaborative meta-analysis of randomized trials. *Eur Heart J.* 2016;37(4): 390−399.

51. Ariotti S, Adamo M, Costa F, et al. Is bare-metal stent implantation still justifiable in high bleeding risk patients undergoing percutaneous coronary intervention?: a pre-specified analysis from the ZEUS trial. *JACC Cardiovasc Interv.* 2016;9(5):426−436.

52. Urban P, Meredith IT, Abizaid A, et al. Polymer-free drug-coated coronary stents in patients at high bleeding risk. *N Engl J Med.* 2015;373(21):2038−2047.

53. Frigoli E, Smits P, Vranckx P, et al. Design and rationale of the management of high bleeding risk patients post bioresorbable polymer coated stent implantation with an abbreviated versus standard DAPT regimen (MASTER DAPT) study. *Am Heart J.* 2019;209:97−105.

54. Costa F, Tijssen JG, Ariotti S, et al. Incremental value of the CRUSADE, ACUITY, and HAS-BLED risk scores for the prediction of hemorrhagic events after coronary stent implantation in patients undergoing long or short duration of dual antiplatelet therapy. *J Ame Heart Assoc.* 2015;4(12):e002524.

55. Costa F, van Klaveren D, James S, et al. Derivation and validation of the predicting bleeding complications in patients undergoing stent implantation and subsequent dual antiplatelet therapy (PRECISE-DAPT) score: a pooled analysis of individual-patient datasets from clinical trials. *Lancet.* 2017;389(10073):1025−1034.

56. Cannon CP, Bhatt DL, Oldgren J, et al. Dual antithrombotic therapy with dabigatran after PCI in atrial fibrillation. *N Engl J Med.* 2017;377(16):1513−1524.

57. Lopes RD, Heizer G, Aronson R, et al. Antithrombotic therapy after acute coronary syndrome or PCI in atrial fibrillation. *N Engl J Med.* 2019;380(16):1509−1524.

58. Gibson CM, Mehran R, Bode C, et al. Prevention of bleeding in patients with atrial fibrillation undergoing PCI. *N Engl J Med.* 2016;375(25):2423−2434.

59. Vranckx P, Valgimigli M, Eckardt L, et al. Edoxaban-based versus vitamin K antagonist-based antithrombotic regimen after successful coronary stenting in patients with atrial fibrillation (ENTRUST-AF PCI): a randomised, open-label, phase 3b trial. *Lancet.* 2019;0(0).

60. Fiedler KA, Maeng M, Mehilli J, et al. Duration of triple therapy in patients requiring oral anticoagulation after drug-eluting stent implantation: the ISAR-TRIPLE trial. *J Am Coll Cardiol.* 2015;65(16):1619−1629.

61. Costa F, Valgimigli M, Steg G, et al. *Triple Antithrombotic Therapy with Warfarin or Dual Therapy with Dabigatran According to the PRECISE-DAPT Score in Patients with Atrial Fibrillation Undergoing Percutaneous Coronary Intervention: Insights from the Re-dual Pci Trial.* Philadelphia: American Heart Association; 2019.

62. Costa F, Vranckx P, Leonardi S, et al. Impact of clinical presentation on ischaemic and bleeding outcomes in patients receiving 6- or 24-month duration of dual-antiplatelet therapy after stent implantation: a pre-specified analysis from the PRODIGY (prolonging dual-antiplatelet treatment after grading stent-induced intimal hyperplasia) trial. *Eur Heart J.* 2015;36(20):1242−1251.

63. Bhatt DL, Fox KA, Hacke W, et al. Clopidogrel and aspirin versus aspirin alone for the prevention of atherothrombotic events. *N Engl J Med.* 2006;354(16):1706−1717.
64. Virmani R, Kolodgie FD, Burke AP, et al. Atherosclerotic plaque progression and vulnerability to rupture: angiogenesis as a source of intraplaque hemorrhage. *Arterioscler Thromb Vasc Biol.* 2005;25(10):2054−2061.
65. Kolodgie FD, Gold HK, Burke AP, et al. Intraplaque hemorrhage and progression of coronary atheroma. *N Engl J Med.* 2003;349(24):2316−2325.
66. Mujaj B, Bos D, Muka T, et al. Antithrombotic treatment is associated with intraplaque haemorrhage in the atherosclerotic carotid artery: a cross-sectional analysis of the Rotterdam Study. *Eur Heart J.* 2018;39(36):3369−3376.
67. Costa F, Van Klaveren D, Feres F, et al. Dual antiplatelet therapy duration based on ischemic and bleeding risks after coronary stenting. *J Am Coll Cardiol.* 2019;73(7): 741−754.

Dual antiplatelet therapy may prevent coronary ischemic events beyond one year—the case for extended treatment

Michael I. Brener, MD [1] **, Sorin J. Brener, MD** [2]

[1]*Fellow, Cardiovascular Medicine, Columbia University Medical Center-NewYork Presbyterian Hospital, New York, NY, United States;* [2]*Professor of Medicine, NewYork Presbyterian-Brooklyn Methodist Hospital, Brooklyn, NY, United States*

Introduction

As you have appreciated from the preceding chapters, atherosclerosis is a complex pathologic process that develops as a consequence of a number of metabolic derangements. An inflammatory milieu affecting the entire coronary tree, compounded by disordered platelet function that manifests throughout the whole circulatory system, predisposes patients to plaque rupture events and progressive coronary artery stenosis. As such, mechanical solutions to this disease, i.e., balloon angioplasty or coronary stents, have inherent limitations, and treatments to address the underlying biology are essential. Pharmacologic platelet inhibition via antagonism of the $P2Y_{12}$ receptor—the target of the thienopyridines clopidogrel and prasugrel, as well as the cyclopentyl triazolopyrimidine, ticagrelor, which together form the second pillar of dual antiplatelet therapy (DAPT), along with aspirin— plays a critical role in modulating the pathobiology of atherosclerotic cardiovascular disease.

While it is clear that DAPT is an important element of the medical management of coronary artery disease (CAD), the duration of DAPT following percutaneous coronary intervention (PCI) and the optimal agent to use has been a subject of controversy and the focus of dozens of observational studies and randomized clinical trials. Herein, we review the evidence basis for "prolonged DAPT," which is defined in most trials as beyond 12 months of guideline-directed therapy. In selected patients who are vulnerable to recurrent ischemic complications but at low risk for bleeding, extended treatment with DAPT may significantly reduce the morbidity associated with atherosclerotic disease.

Rationale for prolonged dual antiplatelet therapy

The reasoning for prolonged antiplatelet therapy in patients with CAD stemmed largely from two critical findings borne from observational data. First is that patients receiving drug-eluting stents (DESs) are at increased risk for developing late (i.e., more than 1 year after stent implantation) stent thrombosis, and second is that patients with CAD remain at risk for plaque rupture events in lesions in the coronary tree that were not treated mechanically by a stent. These findings, particularly when taken in the context of a number of cellular and molecular studies highlighting the protective effects of antiplatelet therapy,[1–4] created the necessary momentum to drive larger studies testing the benefit of prolonged DAPT.

Two seminal observational studies conducted nearly a decade and a half ago set the stage for heightened concern of late in-stent thrombosis associated with DESs. The BASKET-LATE (Basel Stent Kosten-Effektivitäts Trial - LAte Thrombotic Events Trial) study followed 746 patients enrolled in the BASKET randomized study, which assigned patients to either a DES or bare metal stent (BMS), after they discontinued antiplatelet therapy.[5] The study identified a 4.9% incidence of late in-stent thrombosis in patients who received DESs, compared with 1.9% in patients who received BMS. The LAST (Late Angiographic Stent Thrombosis) study raised further alarm by identifying in-stent thrombosis even among patients who were stably on aspirin monotherapy.[6] These studies established that, while rare, in-stent thrombosis long after DES implantation (particularly with first-generation DES) occurred and appeared to coincide with antiplatelet therapy cessation or de-escalation.

The second justification for prolonged DAPT emerged in 2011, when Stone and colleagues[7] released a groundbreaking report from the PROSPECT (Providing Regional Observations to Study Predictors of Events in the Coronary Tree) study, which was a prospective investigation of 697 patients who underwent PCI. Intravascular ultrasound was performed in all three major coronary vessels at the time of PCI, and clinical outcomes during a median of 3.4 years of follow-up were recorded. The investigators identified an overall 20.4% incidence of major adverse cardiovascular events (MACE) during the follow-up period, and importantly, a nearly equal split between adjudicated events caused by the culprit lesion versus events caused by nonculprit lesions (12.9% vs. 11.6%; see Fig. 8.1). Moreover, the nonculprit lesions were usually angiographically mild at the time of the initial PCI, implying that future ischemic events were both hard to predict and often unlikely to be treated by stents.

These data made clear that patients undergoing PCI were at risk for stent-related complications as well as events related to unstable CAD, and it required only a small leap in logic to assume that more potent antiplatelet therapy could potentially mitigate these adverse events. In an analysis of 4666 patients, Eisenstein and colleagues[8] demonstrated that prolonged DAPT with clopidogrel was associated with improved outcomes, but only in the subset of patients treated with DESs. Among patients who were event-free at 12 months, prolonged clopidogrel use was associated with a lower

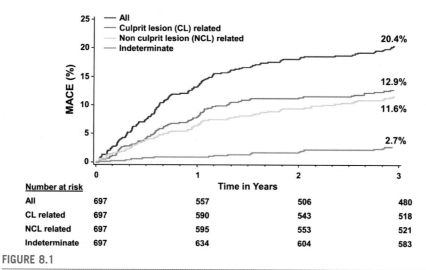

FIGURE 8.1

Kaplan-Meier estimates of 3-year event rates for major adverse cardiovascular outcomes. Note the parity between events adjudicated to be originating from the site of previously treated lesions versus plaque-rupture events that originated elsewhere in the coronary tree. *MACE*, major adverse cardiovascular events.

Adapted from Stone GW, Maehara A, Lansky AJ et al. A prospective natural-history study of coronary athero-sclerosis. N Engl J Med. *2011;364(3):226–235.*

risk of death and myocardial infarction (MI) at 24 months (0% with clopidogrel, 4.5% without clopidogrel, $P < .001$). These studies provided the needed equipoise to test the concept of prolonged DAPT in randomized clinical trials, which are reviewed in the next section.

Randomized trials assessing prolonged dual antiplatelet therapy

Table 8.1 summarizes the major randomized clinical trials to assess the efficacy and safety of extended DAPT. The first such study evaluating this hypothesis that prolonged DAPT may improve ischemic outcomes emerged from South Korea with the combined publication of findings from the REAL-LATE (Correlation of Clopidogrel Therapy Discontinuation in Real-World Patients Treated with Drug-Eluting Stent Implantation and Late Coronary Arterial Thrombotic Events) and ZEST-LATE (Evaluation of the Long-Term Safety after Zotarolimus-Eluting Stent, Sirolimus-Eluting Stent, or Paclitaxel-Eluting Stent Implantation for Coronary Lesions — Late Coronary Arterial Thrombotic Events) studies.[9] A total of 2701 patients who underwent PCI and tolerated DAPT for a minimum of 12 months without ischemic or bleeding complications were randomized to continue DAPT

Table 8.1 Randomized clinical trials investigating extended dual antiplatelet therapy.

Study name	Year	N	DAPT duration (months)	P2Y12 inhibitor	ACS (%)	Primary endpoint and result		Bleeding measure and result	
REAL-LATE ZEST-LATE	2010	2701	12 versus 24	Clopidogrel	62	CV death or MI	↕	TIMI major bleeding	↕
PRODIGY	2012	2013	≤6 versus 24	Clopidogrel	74	All-cause death, MI, or stroke at 2 years	↕	BARC class 2, 3, and 5 bleeding	←
ARCTIC Interruption	2014	1259	12 versus 18	Clopidogrel (91%) Prasugrel (9%)	26	All-cause mortality, MI, stroke, stent thrombosis, urgent revascularization	↕	STEEPLE major bleeding	←
DAPT	2014	9961	12 versus 30	Clopidogrel (66%) Prasugrel (34%)	43	(1) All-cause death, MI, or stroke at 30 months; (2) in-stent thrombosis	→	BARC class 2, 3, and 5 bleeding	←
DES-LATE	2014	5045	12 versus 36	Clopidogrel	61	CV death, MI, and stroke at 2 years	↕	TIMI major bleeding	↕
ITALIC	2015	2031	6 versus 24	Clopidogrel (98%) Prasugrel (2%)	23	All-cause death, MI, stroke, target vessel revascularization	↕	TIMI major bleeding	a
PEGASUS-TIMI 54	2015	21,162	N/A[b]	Ticagrelor	100	CV death, MI, and stroke at 3 years	→	TIMI major bleeding	←
OPTIDUAL	2016	1385	12 versus 36	Clopidogrel	36	All-cause death, MI, and stroke	↕	ISTH major bleeding	↕

ACS, acute coronary syndrome; BARC, Bleeding Academic Research Consortium; CV, cardiovascular; DAPT, dual antiplatelet therapy; ISTH, International Society of Thrombosis and Hemorrhage; MI, myocardial infarction; STEEPLE, Safety and Efficacy Of Enoxaparin In Percutaneous Coronary Intervention Patients; TIMI, Thrombolysis in Myocardial Infarction.

[a] Insufficient bleeding events for statistical analysis.

[b] Patients were enrolled 1–3 years after an MI and randomized to ticagrelor versus placebo with a median of 33 months of follow-up. Prerandomization DAPT exposure was not uniform.

Adapted from Howard, CE et al. Extended duration of dual-antiplatelet therapy after percutaneous coronary intervention: how long is too long? J Am Heart Assoc. 2019 Oct 15;8(20);e012639.

or aspirin monotherapy. Owing to a very low event rate in the pooled trials, the analysis was underpowered, and thus no significant differences emerged with respect to the composite endpoint of cardiovascular mortality and MI at 24 months (1.8% with DAPT vs. 1.2% with aspirin monotherapy; hazard ratio [HR] 1.65, 95% confidence interval [CI], 0.80–3.36, $P = .17$).

These data were echoed in 2012, when the PRODIGY (Prolonging Dual Antiplatelet Treatment After Grading Stent-Induced Intimal Hyperplasia Study) investigators reported a study of 2013 patients undergoing PCI randomized to either 6 or 24 months of DAPT in addition to one of the four stent types (BMS, zotarolimus-DES, everolimus-DES, or paclitaxel-DES).[10] Unlike the REAL-LATE and ZEST-LATE studies, PRODIGY was adequately powered. With respect to the primary endpoint of death, MI, or cerebrovascular accident, there was no difference between 6 and 24 months of DAPT, even when in-stent thrombosis was added to the definition of the composite outcome. One notable subgroup, i.e., patients with peripheral arterial disease, benefited with prolonged DAPT. However, patients with stable CAD appeared to suffer the most from prolonged DAPT, experiencing a fivefold increase in Bleeding Academic Research Consortium (BARC) class 2, 3, and 5 events.[11] This observation was fortified by the 2014 publication of the ARCTIC-Interruption (Assessment by a Double Randomization of a Conventional antiplatelet Strategy Versus a Monitoring-Guided Strategy for Drug-Eluting Stent Implantation and of Treatment Interruption Versus Continuation One Year After Stenting-Interruption) study, which recruited patients at low risk for recurrent ischemic events and demonstrated no improvement in the primary endpoint of all-cause mortality, MI, stroke, urgent target vessel revascularization, and stent thrombosis, but an increase in major bleeding events.[12]

These findings set the stage for the landmark Dual Antiplatelet Therapy (DAPT) study, which evaluated 9961 patients treated with DES-PCI who tolerated 12 months of DAPT and randomized them to DAPT cessation (aspirin monotherapy) versus continuation for a total of 30 months of treatment.[13] The study was designed to be powered for three outcomes at 30 months after PCI, including major adverse cardiovascular and cerebrovascular events (MACCE), stent thrombosis, and moderate or severe bleeding using the GUSTO (Global Use of Strategies to Open Occluded Coronary Arteries) scale. While clopidogrel was the dominant $P2Y_{12}$ inhibitor used, 34% of patients received prasugrel as part of their DAPT regimen.

MACCE were reduced in patients who received prolonged DAPT relative to aspirin (4.3% vs. 5.9%, HR 0.71, 95% CI, 0.59–0.85, $P < .001$). Stent thrombosis also occurred less frequently (0.4% vs. 1.4%, $P < .001$). These positive findings were tempered, however, by increased bleeding (GUSTO moderate-severe bleeding: 2.5% vs. 1.6%, $P = .001$; BARC class 2, 3, and 5 bleeding: 5.6% vs. 2.9%, $P < .001$). These bleeding events were associated with mortality more frequently in the prolonged DAPT arm (11 deaths attributed to bleeding vs. 3 deaths in the placebo arm). Furthermore, overall mortality was increased in the prolonged DAPT arm relative to aspirin (2.0% vs. 1.5%, HR 1.36, 95%, CI 1.00–1.85, $P = .05$), although this finding was driven largely by noncardiovascular mortality.

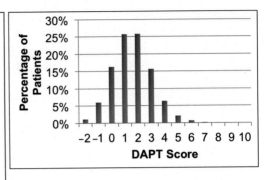

Variable	Points
Patient Characteristic	
Age	
≥75	−2
65 to <75	−1
<65	0
Diabetes Mellitus	1
Current Cigarette Smoker	1
Prior PCI or Prior MI	1
CHF or LVEF <30%	2
Index Procedure Characteristic	
MI at Presentation	1
Vein Graft PCI	2
Stent Diameter <3mm	1

FIGURE 8.2

The dual antiplatelet therapy (DAPT) score, ranging from −2 to 10, is composed of the patient's age, a history of cigarette smoking, diabetes mellitus, myocardial infarction (MI) at presentation, prior MI or percutaneous coronary intervention (PCI), use of paclitaxel-eluting stents, stent diameter <3 mm, a history of congestive heart failure or ejection fraction <30%, and intervention on a vein graft. A higher score (i.e., ≥2) is associated with greater benefit from prolonged DAPT. *CHF*, congestive heart failure; *LVEF*, left ventricular ejection fraction.

Adapted from Yeh RW, Secemsky EA, Kereiakes DJ et al. Development and validation of a prediction rule for benefit and harm of dual antiplatelet therapy beyond 1 year after percutaneous coronary intervention. JAMA. 2016;315(16):1735–1749.

Recognizing the inherent trade-off of reducing ischemic events while increasing bleeding events associated with DAPT, Yeh and colleagues[14] used the DAPT trial cohort to identify risk factors for recurrent ischemic events and bleeding complications (see Fig. 8.2). This topic is addressed in detail in Chapter 9. Suffice it to note that the score addressed the combined net effect of bleeding and ischemic complications, without addressing the relative importance of each component in predicting the risk of death.

Thus the DAPT trial provided the first robust evidence from a randomized clinical trial to support the practice of prolonged DAPT. Although there was a signal for harm for certain patients, the DAPT score provided an evidence-based strategy of identifying patients with the strongest odds of deriving protection from ischemic events while minimizing the risk of bleeding. Subsequent to the publication of the DAPT trial, a number of other randomized trials have been published and provide further insight into the benefits of prolonged DAPT.

DES-LATE (Optimal Duration of Clopidogrel Therapy With DES to Reduce Late Coronary Arterial Thrombotic Event) was a study that combined a cohort of 2344 patients treated with PCI who tolerated 12 months of DAPT with a cohort of 2701 patients assessed in Park et al.'s analysis of the REAL-LATE and

ZEST-LATE studies. Participants were randomized to aspirin monotherapy versus DAPT for an additional 12 months. The study was an open-label study with limited adherence to clopidogrel (79.4% at 2-year follow-up in the extended DAPT arm) as well as a not-insignificant crossover rate where 8.1% of patients assigned to aspirin monotherapy received clopidogrel. As such, the primary endpoint of cardiovascular mortality, MI, and stroke was not significantly different between both arms (aspirin monotherapy arm 2.4% vs. DAPT arm 2.6%, HR 0.94, 95% CI, 0.66−1.35, $P = .75$). Definite stent thrombosis was also not statistically significantly different (11 patients in aspirin monotherapy vs. 7 patients in DAPT arm, HR 1.59, 95% CI, 0.61−4.09, $P = .34$). At 24 months, Thrombolysis in Myocardial Infarction (TIMI) major bleeding events were also similar between the aspirin monotherapy and DAPT arms (1.1% vs. 1.4%, HR 0.71, 95% CI, 0.42−1.20, $P = .20$).

The ITALIC (Is There A LIfe for DES After Discontinuation of Clopidogrel) and OPTIDUAL (OPTImal DUAL antiplatelet therapy) studies were both plagued by slow recruitment and were terminated prematurely as a result. These methodological barriers limit the conclusions that can be inferred from the data.[15,16] Nevertheless, OPTIDUAL, for example, recruited patients who tolerated DAPT for 12 months following PCI and randomized them to DAPT cessation versus continuation. The study design was unique in so far as the investigators assessed net benefit of extended DAPT with a composite endpoint that included all-cause mortality, MI, stroke, and major bleeding. However, recruitment of patients proceeded slowly, such that only 1398 of the proposed 1966 patients required to achieve adequate statistical power were actually enrolled. The primary outcome was not significantly different between both study arms (extended DAPT 5.8% vs. P2Y$_{12}$ inhibition cessation 7.5%, HR 0.75, 95% CI, 0.50−1.28, $P = .17$).

The early round of randomized trials assessing extended DAPT, therefore, provided equivocal results. Although there were multiple trials that questioned the efficacy of extending DAPT beyond 12 months after PCI, the largest of them all, the DAPT study, was adequately powered and suggested that 30 months of DAPT may be superior to the traditional 12 months after PCI. In particular, specific patient characteristics predicted benefit from DAPT such that subsequent studies targeted vulnerable patient populations to enrich the number of expected recurrent ischemic events and utilized more potent antiplatelet agents.

Moving beyond clopidogrel

In this vein, the PEGASUS-TIMI 54 (Prevention of Cardiovascular Events in Patients with Prior Heart Attack Using Ticagrelor Compared to Placebo on a Background of Aspirin−Thrombolysis in Myocardial Infarction 54) investigators identified 21,162 patients who suffered an MI in the previous 1−3 years and had additional risk factors for recurrent MI, including age $\geq$65 years, diabetes mellitus, multivessel CAD, two or more MIs, and chronic renal dysfunction.[17] Participants were randomized in a 1:1:1 fashion to receive ticagrelor 90 mg twice daily, 60 mg twice daily, or placebo.

The study's primary composite endpoint of cardiovascular mortality, MI, and stroke occurred at 3 years in 9.04% of patients in the placebo group, in 7.85% of patients treated with the 90 mg dose (HR vs. placebo 0.85, 95% CI, 0.75–0.96, $P = .008$), and in 7.77% of patients treated with the 60 mg dose (HR vs. placebo 0.84, 95% CI, 0.74–0.95, $P = .004$). Although TIMI major bleeding was more frequent with ticagrelor, the rates of intracranial or fatal hemorrhage were equivalent. Thus PEGASUS-TIMI 54, the largest randomized trial of extended DAPT, substantiated some of the key findings reported from the DAPT trial and provided robust support for prolonged DAPT with a potent P2Y$_{12}$ inhibitor such as ticagrelor (see Fig. 8.3).

Interestingly, a substudy from PEGASUS-TIMI 54 examining the types of MI that occurred during the study period showed that the incidence of type 4b MI (i.e., stent thrombosis) was not significantly reduced with either dose of ticagrelor (90 mg dose, 0.44% vs. 60 mg dose, 0.54% vs. placebo, 0.57%; pooled ticagrelor doses vs. placebo HR 0.82, 95% CI, 0.55–1.22, $P = .33$).[18] These data suggest that in an era with markedly improved stent technology where stent thrombosis is

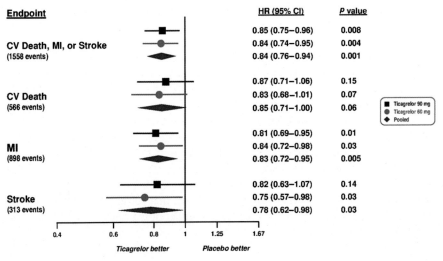

FIGURE 8.3

Ticagrelor at both doses—90 and 60 mg twice daily—showed substantial improvement relative to placebo across a wide range of endpoints, including myocardial infarction (MI) and stroke, in the PEGASUS-TIMI 54 (Prevention of Cardiovascular Events in Patients with Prior Heart Attack Using Ticagrelor Compared to Placebo on a Background of Aspirin—Thrombolysis in Myocardial Infarction 54) study. *CI*, confidence interval; *CV*, cardiovascular; *HR*, hazard ratio.

Adapted from Bonaca MP, Bhatt DL, Cohen M et al. Long-term use of ticagrelor in patients with prior myocardial infarction. N Engl J Med. 2015;372(19):1791–1800.

relatively rare, the benefit of extended DAPT is predominantly realized by a reduction in spontaneous MI.

High-risk patient populations

Although the aforementioned trials recruited patients with CAD, there is significant heterogeneity in the nature of this complex disease, as well as the way it is treated, and the clinical outcomes patients experience following these treatments. This variation must be considered when clinicians recommend the type of antiplatelet agent and the duration of DAPT, prompting many to it an opportunity to practice "personalized medicine" by taking into account specific patient characteristics and treatment details.[19]

For example, patients with a history of plaque-rupture MI appear to benefit more from continued DAPT than patients with stable, symptomatic CAD treated with PCI. This finding has been demonstrated and replicated in a number of the key trials discussed to this point. In DAPT for example, 30.7% of patients enrolled had presented with an MI, and this subgroup appeared to experience an exaggerated treatment effect of extended DAPT with respect to MACCE (with MI 3.9% vs. 6.8%, $P < .001$ compared to those without MI 4.4% vs. 5.3%, $P - .08$; P-interaction $- .03$).[20] A large meta-analysis of 33,435 patients robustly confirmed these data, demonstrating a reduction in recurrent MI (relative risk [RR] 0.70, 95% CI, 0.55−0.88, $P - .003$), stent thrombosis (RR 0.50, 95% CI, 0.28−0.89, $P = .02$), and cardiovascular death (RR 0.85, 95% CI, 0.74−0.98, $P = .03$).[21] The finding of increased noncardiovascular mortality seen in the DAPT trial was not evident in this meta-analysis (RR 1.03, 95% CI, 0.86−1.23, $P = .76$), and there was no increased risk of fatal bleeding with extended DAPT (RR 0.91, 95% CI, 0.53−1.58, $P = .75$).

A gradient of risk for stent thrombosis exists depending on the complexity of PCI, i.e., the number of stents implanted, their location, their size, etc., just as the risk for recurrent MI exists in patients with CAD. In a meta-analysis of 1680 patients who underwent complex PCI, defined by risk enhancers such as three-vessel or lesion PCI, PCI with three or more stents implanted or two stents in a bifurcation lesion, total stent length ≥ 60 mm, or a chronic total occlusion, prolonged DAPT was preferable to short-duration DAPT (3−6 months).[22] MACE occurred in 4.1% of patients undergoing complex PCI treated with prolonged DAPT versus 6.8% of patients treated with short DAPT (HR 0.56, 95% CI, 0.35−0.89, P-interaction compared with noncomplex PCI $= 0.01$). As the number of risk enhancers increased with rising PCI complexity, the magnitude of benefit from prolonged DAPT increased as well.

Current state of guidelines

Integrating insights from the data discussed earlier to allow patients to make an informed decision regarding optimal DAPT duration remains a challenge. However, the trials discussed influenced the current guideline recommendations, which provide a useful framework to personalize DAPT decision-making for each patient. The two major sources of guidelines—the joint American Heart Association (AHA)/American College of Cardiology (ACC) and the European Society of Cardiology Guidelines—cautiously recommend prolonged DAPT under certain circumstances (see Fig. 8.4 for head-to-head comparison[23]). The AHA/ACC guidelines, updated in 2016, provide a Class IIb-LOE (level of evidence) A recommendation for prolonged DAPT for patients with acute coronary syndromes and not at high bleeding risk who underwent PCI.[24] The ESC guidelines, updated in 2017, provide a similar recommendation (Class IIb-LOE B). In the setting of stable CAD, prolonged DAPT earned a Class IIa-LOE A recommendation in patients

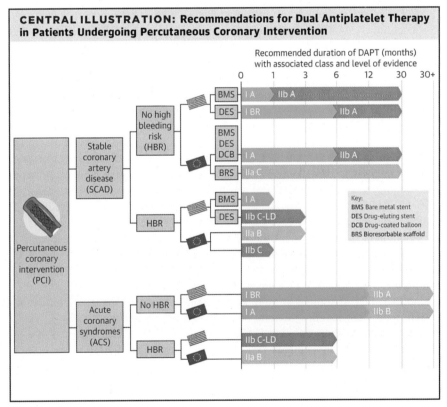

FIGURE 8.4

A comparison of the American and European societal guidelines and recommendations regarding the duration of dual antiplatelet therapy (DAPT).

Reproduced with permission from Capodanno D, Alfonso F, Levine GN, Valgimigli M, Angiolillo DJ. ACC/AHA versus ESC guidelines on dual antiplatelet therapy: JACC guideline comparison. J Am Coll Cardiol. 2018;72(23 Pt A):2915–2931.

not at high bleeding risk and patients may continue DAPT as far out as 30 months after PCI in both sets of guidelines.[25]

The weak recommendation for extended DAPT, despite a number of well-conducted randomized trials demonstrating improved ischemic outcomes with prolonged therapy, stems from three important caveats to the previous discussion. First of all, the magnitude of treatment effect with respect to recurrent MI is relatively small. Udell and colleagues,[21] in a large meta-analysis of patients with MI treated with DAPT, for example, showed that 120 patients needed to receive extended DAPT to prevent one MI. This point is all the more relevant for stent thrombosis, which is a rare complication that is becoming even more rare with improvements in stent technology. Second, the benefits of extended DAPT cannot be separated from the infrequent, but nonetheless present, harms related to the increased bleeding seen with prolonged DAPT. Finally, and most importantly, prolonged DAPT has not demonstrated a beneficial effect on overall mortality in any of the randomized trials discussed thus far. Therefore extending DAPT may be appropriate for certain patients and the decision to prescribe prolonged courses of DAPT should be made on a case-by-case basis.

Conclusion

DAPT serves a critical role in the mitigation of adverse ischemic events in patients with CAD by decreasing the risk of stent thrombosis and spontaneous MI from plaque-rupture events throughout the coronary circulation. For patients with risk factors for recurrent ischemic events—those with a predisposition to or a history of plaque-rupture events or those who underwent complex PCI—extending DAPT duration can reduce cardiovascular mortality, MI, and stent complications. However, bleeding risk must be continuously evaluated, and DAPT should only be continued under circumstances when bleeding risk is low.[26] The "optimal" antiplatelet regimen may evolve as we learn more about the pathogenesis and risk factors for ischemic events and bleeding complications, but some patients at elevated risk for ischemic events will certainly benefit from more aggressive antiplatelet therapies. As discussed in Chapter 9, it may be aspirin that needs to be removed from DAPT.

References

1. Azar RR, Kassab R, Zoghbi A, et al. Effects of clopidogrel on soluble CD40 ligand and on high-sensitivity C-reactive protein in patients with stable coronary artery disease. *Am Heart J*. 2006;151(2):521.e521−521.e524.
2. Afek A, Kogan E, Maysel-Auslender S, et al. Clopidogrel attenuates atheroma formation and induces a stable plaque phenotype in apolipoprotein E knockout mice. *Microvasc Res*. 2009;77(3):364−369.

3. Waksman R, Pakala R, Roy P, et al. Effect of clopidogrel on neointimal formation and inflammation in balloon-denuded and radiated hypercholesterolemic rabbit iliac arteries. *J Interv Cardiol.* 2008;21(2):122–128.
4. Vivekananthan DP, Bhatt DL, Chew DP, et al. Effect of clopidogrel pretreatment on periprocedural rise in C-reactive protein after percutaneous coronary intervention. *Am J Cardiol.* 2004;94(3):358–360.
5. Pfisterer M, Brunner-La Rocca HP, Buser PT, et al. Late clinical events after clopidogrel discontinuation may limit the benefit of drug-eluting stents: an observational study of drug-eluting versus bare-metal stents. *J Am Coll Cardiol.* 2006;48(12):2584–2591.
6. Ong ATL, McFadden EP, Regar E, de Jaegere PPT, van Domburg RT, Serruys PW. Late angiographic stent thrombosis (LAST) events with drug-eluting stents. *J Am Coll Cardiol.* 2005;45(12):2088–2092.
7. Stone GW, Maehara A, Lansky AJ, et al. A prospective natural-history study of coronary atherosclerosis. *N Engl J Med.* 2011;364(3):226–235.
8. Eisenstein EL, Anstrom KJ, Kong DF, et al. Clopidogrel use and long-term clinical outcomes after drug-eluting stent implantation. *Jama.* 2007;297(2):159–168.
9. Park SJ, Park DW, Kim YH, et al. Duration of dual antiplatelet therapy after implantation of drug-eluting stents. *N Engl J Med.* 2010;362(15):1374–1382.
10. Valgimigli M, Campo G, Monti M, et al. Short- versus long-term duration of dual-antiplatelet therapy after coronary stenting: a randomized multicenter trial. *Circulation.* 2012;125(16):2015–2026.
11. Costa F, Vranckx P, Leonardi S, et al. Impact of clinical presentation on ischaemic and bleeding outcomes in patients receiving 6- or 24-month duration of dual-antiplatelet therapy after stent implantation: a pre-specified analysis from the PRODIGY (Prolonging Dual-Antiplatelet Treatment After Grading Stent-Induced Intimal Hyperplasia) trial. *Eur Heart J.* 2015;36(20):1242–1251.
12. Collet JP, Silvain J, Barthelemy O, et al. Dual-antiplatelet treatment beyond 1 year after drug-eluting stent implantation (ARCTIC-Interruption): a randomised trial. *Lancet.* 2014;384(9954):1577–1585.
13. Mauri L, Kereiakes DJ, Yeh RW, et al. Twelve or 30 months of dual antiplatelet therapy after drug-eluting stents. *N Engl J Med.* 2014;371(23):2155–2166.
14. Yeh RW, Secemsky EA, Kereiakes DJ, et al. Development and validation of a prediction rule for benefit and harm of dual antiplatelet therapy beyond 1 Year after percutaneous coronary intervention. *JAMA.* 2016;315(16):1735–1749.
15. Gilard M, Barragan P, Noryani AAL, et al. 6- versus 24-month dual antiplatelet therapy after implantation of drug-eluting stents in patients nonresistant to aspirin: the randomized, multicenter ITALIC trial. *J Am Coll Cardiol.* 2015;65(8):777–786.
16. Helft G, Steg PG, Le Feuvre C, et al. Stopping or continuing clopidogrel 12 months after drug-eluting stent placement: the OPTIDUAL randomized trial. *Eur Heart J.* 2016;37(4):365–374.
17. Bonaca MP, Bhatt DL, Cohen M, et al. Long-term use of ticagrelor in patients with prior myocardial infarction. *N Engl J Med.* 2015;372(19):1791–1800.
18. Bonaca MP, Wiviott SD, Morrow DA, et al. Reduction in subtypes and sizes of myocardial infarction with ticagrelor in PEGASUS-TIMI 54. *J Am Heart Assoc.* 2018;7(22): e009260.
19. Tahir UA, Yeh RW. Individualizing dual antiplatelet therapy duration after percutaneous coronary intervention: from randomized control trials to personalized medicine. *Expert Rev Cardiovasc Ther.* 2017;15(9):681–693.

20. Yeh RW, Kereiakes DJ, Steg PG, et al. Benefits and risks of extended duration dual antiplatelet therapy after PCI in patients with and without acute myocardial infarction. *J Am Coll Cardiol.* 2015;65(20):2211–2221.

21. Udell JA, Bonaca MP, Collet JP, et al. Long-term dual antiplatelet therapy for secondary prevention of cardiovascular events in the subgroup of patients with previous myocardial infarction: a collaborative meta-analysis of randomized trials. *Eur Heart J.* 2016;37(4): 390–399.

22. Giustino G, Chieffo A, Palmerini T, et al. Efficacy and safety of dual antiplatelet therapy after complex PCI. *J Am Coll Cardiol.* 2016;68(17):1851–1864.

23. Capodanno D, Alfonso F, Levine GN, Valgimigli M, Angiolillo DJ. ACC/AHA versus ESC guidelines on dual antiplatelet therapy: JACC guideline comparison. *J Am Coll Cardiol.* 2018;72(23 Pt A):2915–2931.

24. Levine GN, Bates ER, Bittl JA, et al. 2016 ACC/AHA guideline focused update on duration of dual antiplatelet therapy in patients with coronary artery disease: a report of the American College of Cardiology/American Heart Association Task Force on Clinical Practice Guidelines. *J Am Coll Cardiol.* 2016;68(10):1082–1115.

25. Valgimigli M, Bueno H, Byrne RA, et al. 2017 ESC focused update on dual antiplatelet therapy in coronary artery disease developed in collaboration with EACTS: the Task Force for dual antiplatelet therapy in coronary artery disease of the European Society of Cardiology (ESC) and of the European Association for Cardio-Thoracic Surgery (EACTS). *Eur Heart J.* 2017;39(3):213–260.

26. Urban P, Mehran R, Colleran R, et al. Defining high bleeding risk in patients undergoing percutaneous coronary intervention. *Circulation.* 2019;140(3):240–261.

Clinical risk scores: a tool to understand bleeding and thrombotic risk

Robert Yeh, MD, MSc [1], **Nino Mihatov, MD** [2,3]

[1]*Associate Professor of Medicine, Division of Cardiology, Department of Medicine, Beth Israel Deaconess Medical Center, Harvard Medical School, Boston, MA, United States;* [2]*Research Fellow, Richard A. and Susan F. Smith Center for Outcomes Research, Beth Israel Deaconess Medical Center & Harvard Medical School, Boston, MA, United States;* [3]*Clinical & Research Fellow in Cardiovascular Medicine, Division of Cardiology, Department of Medicine, Massachusetts General Hospital & Harvard Medical School, Boston, MA, United States*

Introduction

Dual antiplatelet therapy (DAPT) remains the cornerstone medical therapy in the management of patients following percutaneous coronary intervention (PCI). DAPT protects against stent thrombosis following initial stent deployment and provides long-term cardiovascular benefit by mitigating both stent and non-stent-related outcomes.[1] This benefit comes at a bleeding cost. DAPT-related bleeding is the most common complication after stent implantation, associated with decreased survival and quality of life.[2,3] The optimal DAPT regimen and duration requires a complex risk assessment of both ischemic and bleeding risk, a risk assessment that is often patient specific.

Clinical risk scores have emerged as a useful tool to extrapolate large clinical trial results to individual patients. Frontline clinicians can utilize an accessible tool to rapidly integrate the clinical characteristics of a particular patient to formulate an assessment of a patient's bleeding and ischemic risk profile as a guide for DAPT decision-making. Three of the most commonly used scores integrated into contemporary guidelines include the DAPT score, the PARIS (Patterns of Non-Adherence to Antiplatelet Regimen in Stented Patients) score, and the PRECISE-DAPT (PREdicting bleeding Complications In patients undergoing Stent implantation and subsEquent Dual Anti Platelet Therapy) score (Table 9.1). Each provides complementary information to predict short-term and long-term risks. Here we review the evolution of these and other risk scores in the management of DAPT following PCI.

Table 9.1 Key derivation characteristics of contemporary risk scores for predicting bleeding and/or ischemic risk with DAPT following PCI.

Score	Risk assessment	Derivation cohort (patients, design)	Predicted time interval of risk	Score components	Ischemic endpoint	Ischemic derivation c-statistic	Bleeding endpoint	Bleeding derivation c-statistic
DAPT[39]	Both bleeding and ischemic events (combined)	11,648 patients in RCT of DAPT duration	12–30 months following PCI and event-free at 12 months	Nine clinical variables (three derived from procedural elements)	MI or definite/probable ST	0.70	Out-of-hospital GUSTO moderate or severe bleeding	0.68
PARIS[53]	Both bleeding and ischemic events (separate)	4190 patients in registry of DAPT strategy and duration	At the time of PCI to 24 months	Six clinical variables in each score (two variables overlap)	MI or definite/probable ST	0.70	Out-of-hospital BARC type 3 or 5 bleeding	0.72
PRECISE-DAPT[56]	Bleeding events only	14,963 patients (eight RCTs)	7 days following PCI to 12 months	Five clinical variables	N/A	N/A	Out-of-hospital TIMI major or minor bleeding	0.73 (95% CI, 0.61 –0.85)

BARC, Bleeding Academic Research Consortium; CI, confidence interval; DAPT, dual antiplatelet therapy; GUSTO, Global Utilization of Streptokinase and Tissue Plasminogen Activator for Occluded Arteries; MI, myocardial infarction; PCI, percutaneous coronary intervention; PLATO, PLATelet inhibition and patient Outcomes; RCT, randomized controlled trial; ST, stent thrombosis; TIMI, Thrombolysis in Myocardial Infarction.

Dual antiplatelet therapy: competing bleeding and thrombotic risks

Stent thrombosis is a well-known complication of intracoronary stents and can present early (0–30 days), late (31 days–1 year), or very late (>1 year).[4,5] Early stent data demonstrated a reduction in cardiovascular events in patients with acute coronary syndrome (ACS) treated with PCI with 3–12 months of DAPT when compared with the prior standard of 4 weeks.[6,7] These observations led to the current guideline recommendations offered by the American College of Cardiology (ACC) and the European Society of Cardiology (ESC) recommending DAPT for 12 months following stenting for ACS and 6 months following stenting for stable ischemic heart disease (SIHD).[8,9]

While stent thrombosis within the first 30 days carries the highest mortality, upwards of 60% of stent thromboses can occur beyond 12 months following stent implantation.[10] Longer duration DAPT not only mitigates these stent-related events but also provides protection against non-stent-related thrombosis. The long-term benefit of DAPT has been established in patients with myocardial infarction (MI) who do not undergo PCI, highlighting its role in mitigating non-stent-related thrombosis.[1,11,12] As post-MI patients are at higher risk for recurrent reinfarction, DAPT following PCI for ACS may achieve a dual role of offering protection against future thrombotic events in a high-risk population.[13]

Extension of DAPT to 30 months reduces stent thrombosis and cardiovascular events when compared to the recommended 12 months.[11] As expected, this comes at a cost of an increased incidence of moderate to severe bleeding. In contrast, shorter duration DAPT has suggested noninferiority to the current 6- to 12-month guideline standard with similar rates of ischemic and bleeding events.[14–17] Many of these studies were underpowered for clinically significant bleeding and ischemic endpoints, relying on a composite endpoint to establish noninferiority. The use of a composite bleeding and ischemic endpoint, the two endpoints expected to move in opposite directions with differing DAPT durations, may ultimately bias these study results toward noninferiority and limit their practice changing generalizability. Moreover, many of these trials were open label, which introduces the risk of observer bias, and had only 12 months of follow-up, which limits the detection of late/very late stent thrombosis events.

In spite of the ischemic benefit, the optimal duration of DAPT remains uncertain in light of the competing bleeding risk. Longer duration DAPT has an established association with increased bleeding events.[11] Large randomized controlled trials have sought to assess variable strategies and durations of antiplatelet therapy in an effort to lend clarity to the controversy.[14,17–19] In parallel, second-generation drug-eluting stents (DESs) sought to leverage improved polymer biocompatibility, thinner struts, and optimized pharmacokinetics of antiproliferative drug delivery to reduce the risk of stent thrombosis and potentially limit the duration of adjunctive DAPT.[20]

Studies testing DAPT duration largely enroll patients at low ischemic and bleeding risk. This limits the generalizability of their conclusions to an all-comer PCI population. The fact remains that a large proportion of patients receiving DESs have higher long-term ischemic risk than bleeding risk and ostensibly stand to benefit from longer DAPT.[21] Some, however, have an increased bleeding risk for whom longer DAPT may be harmful. The identification of these patients at high bleeding risk and the determination of the optimal DAPT strategy is more difficult to derive from the existing trial data.

It has become clearer that there may not be a universal approach to DAPT duration. Tailoring the duration of DAPT to each patient's risk profile becomes important. There is ample room in guideline interpretation to empower clinicians to assess bleeding and ischemic risk to individualize DAPT duration. Clinical risk scores have emerged as one tool to achieve this goal of balancing population level evidence with an individualized patient's risk profile.

The perfect risk score

Large clinical trials provide important but averaged estimates of treatment effect across an entire trial population. Subgroup analyses can serve as one limited strategy to tailor large data to individual patients. Subgroup analyses can help identify some groups (i.e., patient with ACS) with larger net clinical benefit from longer duration DAPT.[22,23] There remain, however, patients within these subgroups who are at simultaneously similar or potentially even increased risk of bleeding, for example, over thrombosis, which may not be captured in a simplified subgroup analysis. Moreover, subgroup analyses that help identify different treatment responses are often underpowered to detect meaningful difference in independent bleeding and ischemic endpoints. Lastly, individual patients can present with risk factors that may be independently associated with increased ischemic and/or bleeding risk that may not be adequately discerned in a prespecified subgroup analysis.

Clinical risk scores have emerged as tools to align findings of large clinical trials with individual patient risk profiles. This has enabled a movement away from a "one-size-fits-all" approach. The goal of an optimal risk score is to predict the net clinical benefit of DAPT for an individual patient. Within this paradigm, the derivation of risk scores often occurs from highly selected clinical trial populations. In order to best discern net clinical benefit, risk scores should seek to account for four specific challenges inherent to discerning DAPT risk and benefit.

First, the ideal score must simultaneously weigh bleeding risk with ischemic risk, a problematic discernment because predictors of long-term bleeding events are also predictors of long-term ischemic events[21] (Fig. 9.1). The ideal score would independently discern bleeding and ischemic risk, permitting an identification of predominant risk for a given patient at that point in time. This becomes particularly important in populations, such as the elderly, who have an established simultaneously increased risk of both bleeding and ischemic complications.[24] A score that predicts

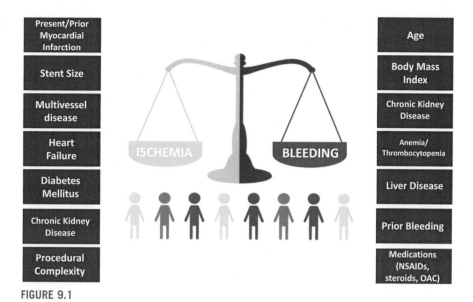

FIGURE 9.1

Representative predictors of ischemic and bleeding events in patients receiving dual antiplatelet therapy. *NSAIDs,* nonsteroidal anti-inflammatory drugs; *OAC,* oral anticoagulation.

one of these events is less informative in comprehensively offering a risk benefit assessment of DAPT continuation.

Second, the ideal risk score must account for the temporal interplay between high upfront ischemic risk with the risk of late bleeding and ischemic events. This interplay becomes more complex because late stent thrombosis results in a much lower case fatality rate than earlier stent thrombosis while late bleeding may be associated with a higher mortality risk.[3,25] Despite this suggestion of late bleeding events carrying higher morbidity and mortality, most early ischemic and bleeding risk algorithms after PCI focused on in-hospital or shorter term event prediction.[26–30] An inherent limitation of any risk assessment is its limited ability to incorporate the dynamic nature of risk over time. The optimal strategy incorporates continual risk reassessment and recognizes that different scores may be required at different time points.

Third, the score must predict events following discharge that are most directly influenced by DAPT strategy, namely, MI, stent thrombosis, and major bleeding. Historical models that have typically focused on in-hospital events have identified procedural characteristics, lesion complexity, stent size, or stent length as predictors of stent thrombosis.[28,29] The optimal score, however, incorporates only those procedural characteristics established to be predictive of longer term events. Early stent thrombosis, for example, does seem to be more related to procedural characteristics than late stent thrombosis, highlighting the benefits of including procedural elements.[28] Complex percutaneous revascularization is similarly associated with increased ischemic events.[31,32] Simultaneously, a risk score should exclude events not known to be affected by variation in DAPT durations, including noncardiac death, cancer-related death, and cancer-related adverse events.

Fourth, the score should depend on clinical variables that are easily identifiable by clinicians and rapidly integrated into a user-friendly interface. The advent of mobile applications has facilitated the expansion of clinical risk score complexity while preserving their usability. While more complex risk scores may improve accuracy, their seamless incorporation into clinical workflows may prove challenging.

Interpreting risk score quality assessment

The developed clinical risk scores are defined and evaluated statistically by their precision and accuracy. Commonly reported c-indices represent the ability of a score to discriminate future cases from noncases. In other words, a c-statistic estimates the probability that a score would assign a randomly selected patient who experiences an event with a higher score (i.e., a score that predicts the event) compared to a patient who does not experience the event.[33,34] Importantly, c-statistics do not discern the probability that patients are categorized correctly. In more pragmatic terms, a c-statistic of 1 indicates a score is perfectly discriminating between cases and noncases. Conversely, a c-statistic of 0.5 implies that the risk score's discriminatory power is comparable to random chance.

Calibration, on the other hand, refers to how accurately a risk score predicts an absolute level of risk. A score may have excellent discriminatory ability as evaluated by the c-statistic but poor calibration. Calibration can be demonstrated by dividing the population into quantiles of predicted risk and plotting the observed event rates for these quantiles. The goodness of fit (GOF) for a risk score tests for a difference between the observed and predicted event rates across quantiles. The Hosmer-Lemeshow χ^2 is a commonly reported, although insensitive, test of GOF. Newer strategies for calibration assessment extend the Hosmer-Lemeshow test to discern how accurately a model aligns expected and predicted event occurrences.[35]

The early applications of risk scores in percutaneous coronary intervention

The early application of risk scores in DAPT leveraged preexisting bleeding scores, derived from patients who did not undergo coronary stenting, to identify individuals at high risk for bleeding. The CRUSADE (Can Rapid Risk Stratification of Unstable Angina Patients Suppress Adverse Outcomes with Early Implementation of the ACC/AHA Guidelines), ACUITY (Acute Catheterization and Urgent Intervention Triage Strategy), and HAS-BLED (Hypertension, Abnormal Renal/Liver Function,

Stroke, Bleeding History or Predisposition, Labile INR, Age, Drugs/Alcohol Use) scores were originally developed for patients with ACS (ACUITY and CRUSADE) or atrial fibrillation (HAS-BLED).[26,30,36] When tested in an all-comer post-PCI population, CRUSADE and ACUITY better predicted major out-of-hospital bleeding than HAS-BLED, including an accurate ability to discern the treatment effect of 24 versus 6 months of DAPT on bleeding risk.[37]

It is important to highlight the several important limitations of these scores. The CRUSADE and ACUITY risk scores were derived from ACS populations, an inherently high-risk group for subsequent ischemic complications. This ischemic risk, however, is left unquantified. Furthermore, the generalizability of these scores to populations with SIHD is unknown and highlights the need to integrate ischemic risk into the assessment of DAPT duration. Their discriminatory power for in-hospital bleeding similarly underscores how all three of the aforementioned scores were originally designed to predict in-hospital and periprocedural events, endpoints less relevant to the net clinical benefit of DAPT,[38] whose early utility is not in dispute. Contemporary risk scores have sought to integrate both ischemic and bleeding risk in the assessment of longer term events.

The DAPT score

To overcome the limitations of preexisting risk scores, the DAPT score was developed from 11648 patients enrolled in the DAPT trial and externally validated in 8136 patients in the Patient-Related Outcomes With Endeavor Versus Cypher Stenting (PROTECT) trial.[39] It was one of the first clinical risk scores to independently and simultaneously predict ischemic events (defined as MI or definite or probable stent thrombosis as specified by the Academic Research Consortium [ARC]) and moderate or severe bleeding events (as defined by Global Use of Strategies to Open Occluded Coronary Arteries [GUSTO] criteria), providing clinicians with a practical tool to guide DAPT duration and subsequently becoming incorporated into contemporary guidelines.[9,40–42]

The DAPT score was developed utilizing separate ischemic and bleeding models to predict the risk of each event from 12 to 30 months of DAPT. Internal validation of these two risk models within the DAPT study cohort demonstrated moderate discrimination for both ischemia (C-statistic: 0.70, 95% confidence interval [CI], 0.68–0.73) and bleeding (C-statistic: 0.68, 95% CI, 0.65–0.72), and both models were well calibrated for both outcomes (GOF: $P = .81$ for ischemia and $P = .34$ for bleeding). External validation in the PROTECT trial demonstrated slight reduction in discrimination of both the ischemia (C-statistic: 0.64, 95% CI, 0.55–0.73) and bleeding models (C-statistic: 0.64, 95% CI, 0.58–0.70).[43]

The absolute difference between predicted ischemic reduction and bleeding increase with longer DAPT (i.e., the predicted effect of longer DAPT on net adverse events) was estimated for each trial participant and then used as an outcome for a linear regression model that identified the predictors that statistically accounted for more than 1% of the observed variation in absolute difference. This led to nine variables (age, history of heart failure, vein graft stenting, MI at presentation, prior MI or PCI, diabetes, stent diameter <3 mm, smoking, and paclitaxel-eluting stent) that produce a score of −2 to +10 (Table 9.2). Notably, some variables with established association with increased bleeding risk, such as anemia or bleeding history, could not be included in the model because they were not collected in the original DAPT study. To ensure that the risk score focused on long-term benefits of DAPT, the DAPT score was developed to be applied only to those patients who tolerated 12 months of DAPT without any major ischemic or bleeding events. A subsequent validation study did demonstrate the score's promise in patients treated for as short as 6 months.[33] Lastly, by virtue of the combination of ischemic and bleeding models, the DAPT score assumes that bleeding and ischemic events are of equal weight.

The DAPT score stratified patients into high-score (DAPT score ≥2) or low-score (DAPT score <2) groups based on the aforementioned six clinical and three procedural variables. High-score patients demonstrated a significant reduction in MI/stent thrombosis and cardiovascular and cerebrovascular events (2.7% vs. 5.7%, $P < .001$, number needed to treat [NNT]: 33) after 30 months of DAPT. There was no significant increase in bleeding with long DAPT duration among patients with high DAPT score (1.8% vs. 1.4%, $P = .26$ and number needed to harm [NNH]: 250). Conversely, low-score patients experienced a significant increase in bleeding (3.0% vs. 1.4%, $P < .001$, NNH: 63) without any ischemic benefit (1.7% vs. 2.3%, $P = .07$, NNT: 167).

Table 9.2 DAPT score variables and associated values.

DAPT score	
Variable	**Value**
Age:	
≥75 years	−2
65–75 years	−1
<65 years	0
Current cigarette smoker	1
Diabetes mellitus	1
MI at presentation	1
Previous PCI or previous MI	1
CHF or LVEF <30%	2
Stent diameter <3 mm	1
Vein graft PCI	2
Paclitaxel-eluting stent	1

A cumulative score <2 indicates low risk, whereas a cumulative score ≥2 indicates high risk.
CHF, congestive heart failure; DAPT, dual antiplatelet therapy; LVEF, left ventricular ejection fraction; MI, myocardial infarction; PCI, percutaneous coronary intervention.

The DAPT score further improved the ability to predict ischemic events in patients with previous MI, who are at higher risk of delayed ischemic events compared to those patients with no MI history.[44] In patients with low DAPT score, continued thienopyridine use was associated with significantly increased incidence of bleeding with similar rates of ischemic events independent of MI history. Similarly, continued thienopyridine use in patients with high DAPT score was associated with a reduction in MI/stent thrombosis (2.7% vs. 6.0%, $P < .001$ any prior MI; 2.6% vs. 5.2%, $P = .002$ no MI), with comparable bleeding rates to placebo, irrespective of prior MI. This highlights the utility of the DAPT score even in the highest risk ACS population.

The DAPT score demonstrates comparable discernment power in more complex PCI. In complex anatomy (patients undergoing intervention involving an unprotected left main, >2 lesions/vessel, lesion length ≥30 mm, bifurcation with side branch ≥2.5 mm, vein bypass graft, or thrombus containing lesion), patients with high DAPT score randomized to longer thienopyridine use had a greater reduction in ischemic events (3.0% vs. 6.1%, $P < .001$) compared with patients with low DAPT score (1.7% vs. 2.3%; $P = .42$; P value comparing risk differences $= 0.03$).[32] In a registry-based analysis of 1142 patients receiving DESs for left main coronary artery bifurcation, a DAPT score ≥2 was a significant independent predictor of major adverse cardiovascular events, which were reduced by prolonged DAPT.[45]

Following the DAPT score's development, several studies have assessed the score's validity across more heterogeneous study populations[33,46–51] (Fig. 9.2). All consistently demonstrated its improved ability to discern bleeding risk. Patients with a low DAPT score had higher cumulative bleeding rates than patients with a high DAPT score. Similarly, all but one validation study[33] demonstrated higher ischemic rates with a high DAPT score compared to a low DAPT score. These results held true with second-generation DESs.[46,49]

The DAPT score is one of the first scores to successfully discern ischemic and bleeding risk independently in patients who tolerated 12 months of DAPT. Its external validation in distinct and heterogeneous patient populations underscores its clinical utility.

The PARIS score

The PARIS registry was a prospective multicenter observational study of patients undergoing PCI with DESs who received DAPT with clopidogrel. It aimed to investigate the influence of different DAPT duration strategies on adverse events with up to 2 years of follow-up.[52] A heterogeneous population of 4190 patients was studied, including patients receiving oral anticoagulation (OAC), to develop separate risk scores for coronary thrombotic events (defined as stent thrombosis or MI) and major bleeding (defined by the Bleeding Academic Research Consortium [BARC] type 3 or 5 bleeding) with external validation performed in the Assessment of Dual

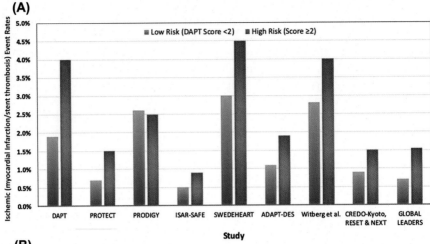

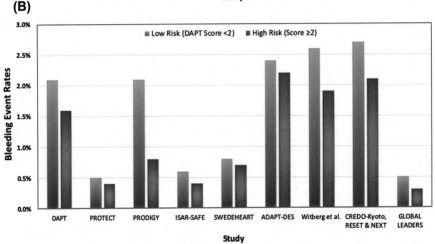

FIGURE 9.2

(A) Ischemic (myocardial infarction and/or stent thrombosis) and (B) bleeding event rates across Dual Antiplatelet Therapy (DAPT) score validation studies stratified by high (≥2) and low (<2) DAPT score.

Antiplatelet Therapy With Drug-Eluting Stents (ADAPT-DES) registry.[53,54] Although only 41% of patients in the derivation cohort and 52% in the original validation cohort presented with ACS, the PARIS score effectively predicted 1-year out-of-hospital bleeding in subsequent validation studies of only patients with ACS.[55]

The PARIS risk score sought to incorporate only those events occurring after discharge and most influenced by DAPT. Thus no procedural characteristics were incorporated. The score included all patients discharged on DAPT, providing a risk assessment at the time of discharge. Unlike the DAPT score, the ischemic and bleeding risk prediction models were not combined to account for differential impacts on morbidity and mortality by any given event. Accordingly, variables predictive of both ischemic and bleeding events were included in each score.

Internal validation of the risk score showed moderate discrimination for ischemic (C-statistic: 0.70) and bleeding (C-statistic: 0.72) events with adequate calibration (GOF: $P = .37$ for ischemic events and $P = .32$ for bleeding events). The external validation in ADAPT-DES, which included contemporary DESs, reaffirmed the moderate level of discrimination for both (C-statistic was 0.65 for ischemic events and 0.64 for bleeding events).

The risk for major bleeding utilized six clinical variables (age, body mass index, smoking status, presence of anemia, impaired creatinine clearance, and triple therapy on discharge, Table 9.3). Patients with high,[8–10] intermediate,[4–7] and low (0–1) bleeding risks were predicted to have a 2-year bleeding event rates of 10.7%, 3.70%, and 1.6%, respectively.

The risk score for coronary thrombotic events included six clinical variables as well (diabetes status, ACS presentation, smoking status, impaired creatinine clearance, prior PCI, and prior CABG), two of which (smoking status and creatinine clearance) overlapped with predictors of major bleeding (Table 9.3). Patients at low, intermediate, and high risks were predicted to have 2-year thrombotic event rates of 1.8%, 3.9%, and 10%, respectively.

When the two models were combined to calculate an absolute risk difference (ARD) as a surrogate for a patient's net clinical benefit from extended DAPT, prolonged DAPT was favored in patients in the higher ischemic risk categories independent of bleeding risk and shorter duration DAPT was favored in patients at higher bleeding risk independent of ischemic risk.

Although the PARIS score does not provide a decision-making algorithm for DAPT duration, it does offer separate risk assessments of bleeding and ischemic risks of DAPT beginning immediately upon discharge from the index PCI hospitalization.

The PRECISE-DAPT score

The PRECISE-DAPT (PREdicting bleeding Complications In patients undergoing Stent implantation and subsEquent Dual Anti Platelet Therapy) score was developed from a pooled dataset of 14963 patients from eight randomized clinical trials treated largely with aspirin and clopidogrel (88% of patients) to predict the risk of out-of-hospital bleeding in individual patients with coronary artery disease treated with stents.[56] In contrast to the DAPT score and similar to the PARIS score, the PRECISE-DAPT score was developed to help guide optimal DAPT duration based on a risk assessment at the time of stent implantation.

Using Cox proportional hazard regression, five clinical variables (age, bleeding history, hemoglobin level, white blood cell count, and creatinine) were identified as predictors of out-of-hospital Thrombolysis in Myocardial Infarction (TIMI) major

Table 9.3 The PARIS (Patterns of Non-Adherence to Antiplatelet Regimen in Stented Patients) ischemic and bleeding risk score variables and associated values.

Ischemic risk score		Bleeding risk score	
Variable	**Value**	**Variable**	**Value**
Age (years)		*Diabetes mellitus*	
<50	0	None	0
50−59	+1	Non-insulin-dependent	+1
60−69	+2	Insulin dependent	+3
70−79	+3	*Acute coronary syndrome*	
≥80	+4	No	0
BMI (kg/m²)		Yes, Tn-negative	+1
<25	+2	Yes, Tn-positive	+2
25−34.9	0	*Currently smoking*	
≥35	+2	Yes	+1
Currently smoking		No	0
Yes	+2	*CrCl <60 mL/min*	
No	0	Present	+2
Anemia		Absent	0
Present	+3	*Prior PCI*	
Absent	0	Yes	+2
CrCl <60 mL/min		No	0
Present	+2	*Prior CABG*	
Absent	0	Yes	+2
Triple therapy on discharge		No	0
Yes	+2		
No	0		

A cumulative ischemic risk score of 0−2 indicates low risk, 3−4 indicates intermediate risk, and ≥5 indicates high risk. A cumulative bleeding risk score of 0−3 indicates low risk, 4−7 indicates intermediate risk, and ≥8 indicates high risk.
ACS, acute coronary syndrome; BMI, body mass index; CABG, coronary artery bypass graft; CAD, coronary artery disease; CrCl, creatinine clearance; PCI, percutaneous coronary intervention; Tn, troponin.

or minor bleeding occurring at least 7 days after the initial procedure.[54] Each variable was assigned a value based on the magnitude of its influence on TIMI minor or major bleeding. A threshold of ≥25 was established as the boundary above which patients experienced clinical harm if treated with a DAPT regimen longer than 3−6 months after stenting. The score was externally validated in the PLATelet inhibition and patient Outcomes (PLATO) trial and the Bern PCI registry.[56]

The PRECISE-DAPT score internal validation showed moderate discrimination with a c-index for out-of-hospital TIMI major or minor bleeding of 0.73 (95% CI, 0.61−0.85) in the derivation cohort, 0.70 (95% CI, 0.65−0.74) in the PLATO trial validation cohort, and 0.66 (95% CI, 0.61−0.71) in the Bern PCI registry validation cohort.

Patients at high bleeding risk (PRECISE-DAPT score ≥25) are exposed to more bleeding with longer DAPT duration (ARD: 2.59%, 95% CI, 0.82−4.34, NNT: 38), whereas patients at low bleeding risk (PRECISE-DAPT score <25) do not show a difference in bleeding with long versus short DAPT (ARD: 0.14%, 95% CI, −0.22 to 0.49, interaction $P = .007$).

In contrast to the DAPT and PARIS scores, the PRECISE-DAPT score was not designed to assess ischemic risk. PRECISE-DAPT, however, suggests that the low-bleeding-score group may derive ischemic benefit without bleeding harm with longer DAPT (ARR: −1.53%, 95% CI, −2.64 to −0.41, NNT: 65). This mirrors the findings of the PARIS score where patients with high bleeding risk demonstrated harm with longer DAPT duration independent of ischemic risk score. When the PRECISE-DAPT score was applied to patients undergoing complex PCI (a previously demonstrated surrogate for increased ischemic risk), the patients benefited from long-term DAPT duration only if features of high bleeding risk by PRECISE-DAPT were absent.[57] This suggests that when assessed independently, the net clinical benefit of DAPT may be disproportionately determined by bleeding risk.

The score was subsequently validated in prospective registry studies and ACS cohorts.[55,58,59] The derivation of the original PRECISE-DAPT score centered on patients treated with aspirin and clopidogrel, with subsequent validation in patients receiving prasugrel or ticagrelor.[60] Utilizing the PRECISE-DAPT score after second-generation DES implantation helped reduce the number of ischemic events at 12 months, particularly in patients with ACS.[61] The 2017 ESC guidelines on DAPT endorse the use of the PRECISE-DAPT score to estimate bleeding risk, recommending 3 months (SIHD) or 6 months (ACS) of DAPT in patients with high bleeding risk (PRECISE-DAPT ≥25).[9]

When PRECISE-DAPT was compared to the PARIS score in ACS subpopulations, both PRECISE-DAPT and PARIS effectively discriminated BARC type 3 or 5 bleeding, a surrogate for major bleeding.[55] Nonetheless, the PARIS score appears superior to the PRECISE-DAPT score in stratifying 1-year out-of-hospital BARC type 2, 3, or 5 bleeding and BARC type 3 or 5 bleeding when indices of net reclassification improvement and integrated discrimination are considered.[55] In patients at high risk with ACS receiving the more potent P2Y12 inhibitors prasugrel and ticagrelor, the PRECISE-DAPT score performed better than the PARIS score in predicting major bleeding.[60] Thus, both scores have a discriminatory role in bleeding risk stratification in the ACS population.

The PRECISE-DAPT offers a powerful assessment of bleeding events following PCI. It does not, however, offer an individual assessment of ischemic risk in those patients undergoing complex PCI or with ACS presentations. Although both the PRECISE-DAPT score and PARIS score suggest that bleeding risk assessment alone predicts harm with prolonged DAPT, there is insufficient corroborating evidence in specific populations, such as the elderly, who are known to be at high risk for both ischemic and bleeding complications.

Applying risk scores to practice

The DAPT, PARIS, and PRECISE-DAPT scores are the three mainstream risk scores described in the current guidelines. All should be seen as complementary tools to provide an individualized risk assessment of bleeding and ischemic risk following PCI (Table 9.4). Each carries strengths and weaknesses resulting from their unique derivations.

Importantly, while some have been derived from randomized controlled data, none has been prospectively tested in the setting of a randomized controlled trial that utilized a risk score to guide DAPT duration. Moreover, the scores are optimally designed for application in patients matched to the inclusion and exclusion criteria of the clinical trials from which they were derived. While the generalizability of the scores beyond their original study cohorts may be limited, the repeated validation of many of the scores across more heterogeneous populations is a reassuring testament to their efficacy.

Both the DAPT and PRECISE-DAPT scores were derived from studies of patients randomized to different DAPT durations, providing some insight into their usefulness in predicting the impact of variable DAPT duration on ischemic and bleeding outcomes. The value of the PARIS bleeding and ischemic risk score to tailor DAPT duration remains unclear, as the original validation cohort did not randomize DAPT duration.

Table 9.4 Decision-making tools and interpretation of contemporary risk scores for the assessment of bleeding and/or ischemic risk in DAPT.

Risk score	Optimal time point for assessment	Risk score interpretation	Decision-making tool
DAPT[39]	1 year of DAPT without bleeding or ischemic events	<2: Low risk (bleeding harm from prolonged DAPT) ≥2: High risk (ischemic benefit from prolonged DAPT)	https://tools.acc.org/DAPTriskapp/
PARIS[53]	At the time of stent implantation	Ischemic risk: low; 0–1, intermediate; 3–4, high; ≥5 Bleeding risk: low; 0–3, intermediate; 4–7, high; ≥8	N/A
PRECISE-DAPT[56]	7 days following stent implantation	>25: High bleeding risk <25: Low-intermediate bleeding risk	www.precidedaptscore.com

DAPT, dual antiplatelet therapy.

As a significant proportion of stent thrombosis cases occurs within the first 30 days of stent implantation, integrating the declining risk of stent thrombosis over time with evolving bleeding risk is challenging.[62] Both the PARIS and PRECISE-DAPT risk scores integrate immediate postprocedural risk with delayed risk. Thus PARIS and PRECISE-DAPT may be most useful postprocedurally to identify those patients in whom short-term bleeding risk is high and for whom shorter duration DAPT should be considered between 0 and 12 months. Furthermore, the PARIS score offers independent assessments of bleeding and ischemic risk. This may effectively incorporate any imbalance in the contribution of ischemic and bleeding risk to the net clinical benefit of longer DAPT. The DAPT score, on the other hand, included patients who had undergone PCI and remained without significant ischemic or bleeding events at 12 months. Thus the DAPT score can offer a reassessment of risk for those patients who tolerate 1 year of uneventful DAPT.

The optimal time point of assessment and reassessment of risk remains unknown and is likely as individualized as the variables that any score uses. Continual evaluation of this dynamic risk remains paramount to mitigate DAPT harm.

Oral anticoagulation

Just over 10% of patients who undergo PCI have an indication for long-term OAC therapy.[63] The optimal combination of anticoagulation and antiplatelet therapy for these patients remains uncertain. Numerous studies have demonstrated that DAPT remains superior to vitamin K antagonists to mitigate stent thrombosis after PCI.[64–66] Nonetheless, OAC therapy is superior to DAPT in the prevention of vascular events, including stroke, in patients with atrial fibrillation.[3] Limited data exist, however, to guide the optimal antithrombotic and antiplatelet strategy that integrates the competing risk of stroke into the bleeding and stent thrombosis risk calculus.

Efforts to validate the PRECISE-DAPT score and PARIS score in patients treated with both OAC and antiplatelet therapy suggest some role for these scores in this population. The PRECISE-DAPT score was better able to identify high-risk bleeding than the PARIS score when utilizing TIMI criteria for major or minor bleeding in the patient population receiving OAC and antiplatelet therapy.[54,67] Nonetheless, both PRECISE-DAPT and PARIS were able to discern patients at high bleeding risk according to BARC 3 bleeding or greater.[68] Thus, these limited data suggest that PRECISE-DAPT and PARIS effectively stratify patients receiving OAC and antiplatelet therapy by bleeding risk.

Studies evaluating the impact of OAC in combination with antiplatelet therapy on ischemic and bleeding outcomes are challenged by the same limitations as the original studies of DAPT strategy and duration. No study to date has integrated a risk score to modulate DAPT strategy in the presence of an OAC. Additionally, more studies are needed to inform what, if any, ischemic benefit is garnered from the presence of OAC that may justify a reduction in antiplatelet therapy duration or composition and to understand the overall combined impact of DAPT and OAC on both bleeding and cardiovascular/cerebrovascular thrombotic risk.

The future of dual antiplatelet therapy risk scores

Risk scores remain one critical resource in the effort to specifically tailor DAPT to a patient's particular risk, minimizing bleeding risk while maximizing ischemic benefit. Although current scores facilitate rapid calculation, increasing score complexity may increase accuracy at the expense of usability. The integration of DAPT risk scores within the electronic health record (EHR) and user-friendly mobile applications ("apps") may allow for the clinical implementation of increasingly complex but more precise risk scores. This has already been successfully demonstrated by the integration of complex atrial fibrillation risk score calculators into the EHR.[69]

The evolution of machine learning has further enhanced the power of the EHR, moving away from traditional population-based risk models to algorithms that do not assume linear relationships between variables and outcomes, thus creating higher order assessment of individualized patient outcomes. Machine learning has suggested greater predictive power than traditional risk stratification methods in ACS and could hold similar predictive power in leveraging the EHR to dynamically predict ischemic and bleeding risk.[70]

Risk scores remain one element in the multifaceted approach to tailor DAPT. More recently, phenotypic and genotypic testing has emerged as an added strategy in tailoring DAPT. Phenotypic assessment of P2Y12 therapy focuses on platelet function testing (PFT). PFT measures platelet function after P2Y12 therapy is introduced and has been correlated to both bleeding and ischemic outcomes.[71] Its integration into clinical practice has suggested a predictive power in future ischemic events.[72] Other studies assessing dose-adjusted clopidogrel in response to PFT, however, have failed to demonstrate a clinical benefit.[73] These mixed results may reflect limited understanding in the expected dose response to platelet reactivity or the inherent variability in platelet function related to dose timing.

Genetic testing may complement PFT to guide DAPT strategy. Genetic testing can further identify patients with variable metabolism of traditional P2Y12 inhibitors. Loss of function of CYP2C19 alleles impairs clopidogrel metabolism, for example, and has been linked to higher risks of cardiovascular events in patients undergoing PCI who receive clopidogrel.[74,75] A genotype-guided strategy for the selection of P2Y12 inhibitor therapy is at least noninferior to standard prasugrel or ticagrelor therapy with regard to thrombotic events; it further leads to lower bleeding.[76] Although the data in support of routine PFT and genetic testing remains mixed, testing can be considered when escalation strategies are desired in patients for whom thrombotic risk outweighs bleeding risk or conversely when de-escalation strategies are desired in patients for whom bleeding risk outweighs thrombotic risk.[77] Future trials, including the ongoing TAILOR-PCI trial, seek to evaluate, in a randomized controlled fashion, the use of salivary genetic testing in an all-comer PCI population to help further inform DAPT decision-making.

On the background of all these developments remain parallel investigations of optimal DAPT regimens. More recent trials have suggested that clopidogrel monotherapy may strike a balance between the bleeding risk of DAPT against the ischemic risk of aspirin monotherapy.[78] Ticagrelor monotherapy, however, maintains a comparable ischemic protection to DAPT without any reduction in bleeding risk.[79] Although more work is needed in understanding the implications of P2Y12 monotherapy following PCI, it may hold promise as an alternative therapy in those patients at high bleeding risk for whom less intensive DAPT is desired.

Conclusion

Despite the significant advancements in stent technology and procedural techniques, DAPT remains the cornerstone medical therapy in the prevention of ischemic complications following PCI. These benefits are derived from protection against stent- and non-stent-related ischemic complications. Protection against these thrombotic events comes at a bleeding cost—a risk-benefit calculus that remains patient centered. The DAPT score, PARIS score, and PRECISE-DAPT score have emerged as the first clinical risk scores that could be utilized to identify those patients who benefit or may be harmed from prolonged DAPT. The intensive effort to optimize the poststent management of antiplatelet therapy will only continue to evolve with the parallel evolution of stent technology, antiplatelet regimens, the utilization of EHR data to advance clinical risk scores, and the advancement of our understanding of P2Y12 pharmacodynamics.

References

1. Yusuf S, Zhao F, Mehta SR, et al. Effects of clopidogrel in addition to aspirin in patients with acute coronary syndromes without ST-segment elevation. *N Engl J Med.* 2001; 345(7):494−502.
2. Amin AP, Bachuwar A, Reid KJ, et al. Nuisance bleeding with prolonged dual antiplatelet therapy after acute myocardial infarction and its impact on health status. *J Am Coll Cardiol.* 2013:2130−2138. https://doi.org/10.1016/j.jacc.2013.02.044.
3. Généreux P, Giustino G, Witzenbichler B, et al. Incidence, predictors, and impact of postdischarge bleeding after percutaneous coronary intervention. *J Am Coll Cardiol.* 2015; 66(9):1036−1045.
4. Montalescot G, Brieger D, Dalby AJ, Park S-J, Mehran R. Duration of dual antiplatelet therapy after coronary stenting. *J Am Coll Cardiol.* 2015:832−847. https://doi.org/10.1016/j.jacc.2015.05.053.
5. Holmes DR, Kereiakes DJ, Garg S, et al. Stent thrombosis. *J Am Coll Cardiol.* 2010: 1357−1365. https://doi.org/10.1016/j.jacc.2010.07.016.
6. Steinhubl SR, Berger PB, Mann 3rd JT, et al. Early and sustained dual oral antiplatelet therapy following percutaneous coronary intervention: a randomized controlled trial. *J Am Med Assoc.* 2002;288(19):2411−2420.
7. Mehta SR, Yusuf S, Peters RJ, et al. Effects of pretreatment with clopidogrel and aspirin followed by long-term therapy in patients undergoing percutaneous coronary intervention: the PCI-CURE study. *Lancet.* 2001;358(9281):527−533.

8. Levine GN, Bates ER, Bittl JA, et al. 2016 ACC/AHA guideline focused update on duration of dual antiplatelet therapy in patients with coronary artery disease. *Circulation*. 2016;134(10):e123—e155.

9. Valgimigli M, Bueno H, Byrne RA, et al. 2017 ESC focused update on dual antiplatelet therapy in coronary artery disease developed in collaboration with EACTS. *Eur Heart J*. 2018;39(3):213—260.

10. Armstrong EJ, Feldman DN, Wang TY, et al. Clinical presentation, management, and outcomes of angiographically documented early, late, and very late stent thrombosis. *JACC Cardiovasc Interv*. 2012;5(2):131—140.

11. Mauri L, Kereiakes DJ, Yeh RW, et al. Twelve or 30 months of dual antiplatelet therapy after drug-eluting stents. *N Engl J Med*. 2014;371(23):2155—2166.

12. Bonaca MP, Bhatt DL, Cohen M, et al. Long-term use of ticagrelor in patients with prior myocardial infarction. *N Engl J Med*. 2015;372(19):1791—1800.

13. Stone GW, Witzenbichler B, Guagliumi G, et al. Heparin plus a glycoprotein IIb/IIIa inhibitor versus bivalirudin monotherapy and paclitaxel-eluting stents versus bare-metal stents in acute myocardial infarction (HORIZONS-AMI): final 3-year results from a multicentre, randomised controlled trial. *Lancet*. 2011;377(9784):2193—2204.

14. Hahn J-Y, Song YB, Oh J-H, et al. 6-month versus 12-month or longer dual antiplatelet therapy after percutaneous coronary intervention in patients with acute coronary syndrome (SMART-DATE): a randomised, open-label, non-inferiority trial. *Lancet*. 2018; 391(10127):1274—1284.

15. Gwon H-C, Hahn J-Y, Park KW, et al. Six-month versus 12-month dual antiplatelet therapy after implantation of drug-eluting stents: the Efficacy of Xience/Promus versus Cypher to Reduce Late Loss after Stenting (EXCELLENT) randomized, multicenter study. *Circulation*. 2012;125(3):505—513.

16. Kim B-K, Hong M-K, Shin D-H, et al. A new strategy for discontinuation of dual antiplatelet therapy: the RESET Trial (REal Safety and Efficacy of 3-month dual antiplatelet Therapy following Endeavor zotarolimus-eluting stent implantation). *J Am Coll Cardiol*. 2012;60(15):1340—1348.

17. Schulz-Schüpke S, Byrne RA, Ten Berg JM, et al. ISAR-SAFE: a randomized, double-blind, placebo-controlled trial of 6 vs. 12 months of clopidogrel therapy after drug-eluting stenting. *Eur Heart J*. 2015;36(20):1252—1263.

18. Vranckx P, Valgimigli M, Jüni P, et al. Ticagrelor plus aspirin for 1 month, followed by ticagrelor monotherapy for 23 months vs aspirin plus clopidogrel or ticagrelor for 12 months, followed by aspirin monotherapy for 12 months after implantation of a drug-eluting stent: a multicentre, open-label, randomised superiority trial. *Lancet*. 2018;392(10151):940—949.

19. Franzone A, Piccolo R, Gargiulo G, et al. Prolonged vs short duration of dual antiplatelet therapy after percutaneous coronary intervention in patients with or without peripheral arterial disease: a subgroup Analysis of the PRODIGY randomized clinical trial. *JAMA Cardiol*. 2016;1(7):795—803.

20. El-Hayek G, Bangalore S, Casso Dominguez A, et al. Meta-analysis of randomized clinical trials comparing biodegradable polymer drug-eluting stent to second-generation durable polymer drug eluting stents. *JACC Cardiovasc Interv*. 2017;10(5):462—473.

21. Matteau A, Yeh RW, Camenzind E, et al. Balancing long-term risks of ischemic and bleeding complications after percutaneous coronary intervention with drug-eluting stents. *Am J Cardiol*. 2015;116(5):686—693.

22. Yeh RW, Kereiakes DJ, Steg PG, et al. Benefits and risks of extended duration dual antiplatelet therapy after PCI in patients with and without acute myocardial infarction. *J Am Coll Cardiol*. 2015;65(20):2211–2221.

23. Udell JA, Bonaca MP, Collet J-P, et al. Long-term dual antiplatelet therapy for secondary prevention of cardiovascular events in the subgroup of patients with previous myocardial infarction: a collaborative meta-analysis of randomized trials. *Eur Heart J*. 2016;37(4): 390–399.

24. Khandelwal D, Goel A, Kumar U, Gulati V, Narang R, Dey AB. Frailty is associated with longer hospital stay and increased mortality in hospitalized older patients. *J Nutr Health Aging*. 2012;16(8):732–735.

25. Waksman R, Kirtane AJ, Torguson R, et al. Correlates and outcomes of late and very late drug-eluting stent thrombosis: results from DESERT (International Drug-Eluting Stent Event Registry of Thrombosis). *JACC Cardiovasc Interv*. 2014;7(10):1093–1102.

26. Mehran R, Pocock SJ, Nikolsky E, et al. A risk score to predict bleeding in patients with acute coronary syndromes. *J Am Coll Cardiol*. 2010;55(23):2556–2566.

27. Halkin A, Singh M, Nikolsky E, et al. Prediction of mortality after primary percutaneous coronary intervention for acute myocardial infarction: the CADILLAC risk score. *J Am Coll Cardiol*. 2005;45(9):1397–1405.

28. Dangas GD, Claessen BE, Mehran R, et al. Development and validation of a stent thrombosis risk score in patients with acute coronary syndromes. *JACC Cardiovasc Interv*. 2012:1097–1105. https://doi.org/10.1016/j.jcin.2012.07.012.

29. Palmerini T, Genereux P, Caixeta A, et al. A new score for risk stratification of patients with acute coronary syndromes undergoing percutaneous coronary intervention: the ACUITY-PCI (acute catheterization and urgent intervention triage strategy–percutaneous coronary intervention) risk score. *JACC Cardiovasc Interv*. 2012;5(11): 1108–1116. https://doi.org/10.1016/j.jcin.2012.07.011.

30. Subherwal S, Bach RG, Chen AY, et al. Baseline risk of major bleeding in non–ST-segment–elevation myocardial infarction. *Circulation*. 2009:1873–1882. https://doi.org/10.1161/circulationaha.108.828541.

31. Valgimigli M, Costa F, Byrne R, Haude M, Baumbach A, Windecker S. Dual antiplatelet therapy duration after coronary stenting in clinical practice: results of an EAPCI survey. *EuroIntervention*. 2015:68–74. https://doi.org/10.4244/eijv11i1a11.

32. Yeh RW, Kereiakes DJ, Steg PG, et al. Lesion complexity and outcomes of extended dual antiplatelet therapy after percutaneous coronary intervention. *J Am Coll Cardiol*. 2017; 70(18):2213–2223.

33. Piccolo R, Gargiulo G, Franzone A, et al. Use of the dual-antiplatelet therapy score to guide treatment duration after percutaneous coronary intervention. *Ann Intern Med*. 2017;167(1):17–25.

34. Lloyd-Jones DM. Cardiovascular risk prediction: basic concepts, current status, and future directions. *Circulation*. 2010;121(15):1768–1777.

35. Demler OV, Paynter NP, Cook NR. Tests of calibration and goodness-of-fit in the survival setting. *Stat Med*. 2015;34(10):1659–1680.

36. Pisters R, Lane DA, Nieuwlaat R, de Vos CB, Crijns HJGM, Lip GYH. A novel user-friendly score (HAS-BLED) to assess 1-year risk of major bleeding in patients with atrial fibrillation: the Euro Heart Survey. *Chest*. 2010;138(5):1093–1100.

37. Costa F, Tijssen JG, Ariotti S, et al. Incremental value of the CRUSADE, ACUITY, and HAS-BLED risk scores for the prediction of hemorrhagic events after coronary stent

implantation in patients undergoing long or short duration of dual antiplatelet therapy. *J Am Heart Assoc.* 2015. https://doi.org/10.1161/jaha.115.002524.

38. Hsieh M-J, Wang C-C, Chen C-C, Wang C-L, Wu L-S, Hsieh I-C. HAS-BLED score predicts risk of in-hospital major bleeding in patients with acute non-ST segment elevation myocardial infarction. *Thromb Res.* 2015;136(4):775−780.

39. Yeh RW, Secemsky EA, Kereiakes DJ, et al. Development and validation of a prediction rule for benefit and harm of dual antiplatelet therapy beyond 1 year after percutaneous coronary intervention. *J Am Med Assoc.* 2016;315(16):1735−1749.

40. Levine GN, Bates ER, Bittl JA, et al. Focused Update Writing Group, 2016 ACC/AHA Guideline focused update on duration of dual antiplatelet therapy in patients with coronary artery disease. *J Am Coll Cardiol.* 2016;68(10):1082−1115.

41. Cutlip DE, Windecker S, Mehran R, et al. Clinical end points in coronary stent trials: a case for standardized definitions. *Circulation.* 2007;115(17):2344−2351.

42. GUSTO investigators. An international randomized trial comparing four thrombolytic strategies for acute myocardial infarction. *N Engl J Med.* 1993;329(10):673−682.

43. Camenzind E, Wijns W, Mauri L, et al. Stent thrombosis and major clinical events at 3 years after zotarolimus-eluting or sirolimus-eluting coronary stent implantation: a randomised, multicentre, open-label, controlled trial. *Lancet.* 2012;380(9851):1396−1405.

44. Kereiakes DJ, Yeh RW, Massaro JM, et al. DAPT score utility for risk prediction in patients with or without previous myocardial infarction. *J Am Coll Cardiol.* 2016;67(21):2492−2502.

45. Cho S, Kim J-S, Kang TS, et al. Long-term efficacy of extended dual antiplatelet therapy after left main coronary artery bifurcation stenting. *Am J Cardiol.* 2019. https://doi.org/10.1016/j.amjcard.2019.10.046.

46. Ueda P, Jernberg T, James S, et al. External validation of the DAPT score in a nationwide population. *J Am Coll Cardiol.* 2018;72(10):1069−1078.

47. Brener SJ, Kirtane AJ, Rinaldi MJ, et al. Prediction of ischemic and bleeding events using the dual antiplatelet therapy score in an unrestricted percutaneous coronary intervention population: analysis from the ADAPT-DES registry. *Circ Cardiovasc Interv.* 2018;11(10):e006853.

48. Harada Y, Michel J, Lohaus R, et al. Validation of the DAPT score in patients randomized to 6 or 12 months clopidogrel after predominantly second-generation drug-eluting stents. *Thromb Haemostasis.* 2017;117(10):1989−1999.

49. Yoshikawa Y, Shiomi H, Watanabe H, et al. Validating utility of dual antiplatelet therapy score in a large pooled cohort from 3 Japanese percutaneous coronary intervention studies. *Circulation.* 2018;137(6):551−562.

50. Witberg G, Zusman O, Bental T, et al. Validation of the DAPT score in real-world patients undergoing coronary stent implantation. *Int J Cardiol.* 2019. https://doi.org/10.1016/j.ijcard.2019.08.044.

51. Chichareon P, Modolo R, Kawashima H, et al. DAPT score and the impact of ticagrelor monotherapy during the second year after PCI. *JACC Cardiovasc Interv.* 2020;13(5):634−646. https://doi.org/10.1016/j.jcin.2019.12.018.

52. Mehran R, Baber U, Steg PG, et al. Cessation of dual antiplatelet treatment and cardiac events after percutaneous coronary intervention (PARIS): 2 year results from a prospective observational study. *Lancet.* 2013;382(9906):1714−1722.

53. Baber U, Mehran R, Giustino G, et al. Coronary thrombosis and major bleeding after PCI with drug-eluting stents: risk scores from PARIS. *J Am Coll Cardiol.* 2016;67(19):2224−2234.

54. Mehran R, Rao SV, Bhatt DL, et al. Standardized bleeding definitions for cardiovascular clinical trials: a consensus report from the Bleeding Academic Research Consortium. *Circulation*. 2011;123(23):2736−2747. https://doi.org/10.1161/circulationaha.110.009449.

55. Abu-Assi E, Raposeiras-Roubin S, Cobas-Paz R, et al. Assessing the performance of the PRECISE-DAPT and PARIS risk scores for predicting one-year out-of-hospital bleeding in acute coronary syndrome patients. *EuroIntervention*. 2018;13(16):1914−1922.

56. Costa F, van Klaveren D, James S, et al. Derivation and validation of the predicting bleeding complications in patients undergoing stent implantation and subsequent dual antiplatelet therapy (PRECISE-DAPT) score: a pooled analysis of individual-patient datasets from clinical trials. *Lancet*. 2017:1025−1034. https://doi.org/10.1016/s0140-6736(17)30397-5.

57. Costa F, Van Klaveren D, Feres F, et al. Dual antiplatelet therapy duration based on ischemic and bleeding risks after coronary stenting. *J Am Coll Cardiol*. 2019;73(7): 741−754.

58. Choi SY, Kim MH, Cho Y-R, et al. Performance of PRECISE-DAPT score for predicting bleeding complication during dual antiplatelet therapy. *Circ Cardiovasc Interv*. 2018; 11(12):e006837.

59. Alba AC, Guyatt G. In patients receiving DAPT after coronary stents, the PRECISE-DAPT score predicted bleeding moderately well. *Ann Intern Med*. 2017:JC11.

60. Bianco M, D'ascenzo F, Raposeiras Roubin S, et al. Comparative external validation of the PRECISE-DAPT and PARIS risk scores in 4424 acute coronary syndrome patients treated with prasugrel or ticagrelor. *Int J Cardiol*. 2019. https://doi.org/10.1016/j.ijcard.2019.11.132.

61. Jang J-Y, Lee B-K, Kim J-S, et al. Efficacy and safety of guidelinc-rccommended risk score-directed dual antiplatelet therapy after 2nd-generation drug-eluting stents. *Circ J*. 2019. https://doi.org/10.1253/circj.CJ-19-0667.

62. Palmerini T, Kirtane AJ, Serruys PW, et al. Stent thrombosis with everolimus-eluting stents: meta-analysis of comparative randomized controlled trials. *Circ Cardiovasc Interv*. 2012;5(3):357−364.

63. Secemsky EA, Butala NM, Kartoun U, et al. Use of chronic oral anticoagulation and associated outcomes among patients undergoing percutaneous coronary intervention. *J Am Heart Assoc*. 2016;5(10). https://doi.org/10.1161/JAHA.116.004310.

64. Leon MB, Baim DS, Popma JJ, et al. A clinical trial comparing three antithrombotic-drug regimens after coronary-artery stenting. *N Engl J Med*. 1998:1665−1671. https://doi.org/10.1056/nejm199812033392303.

65. Bertrand ME, Legrand V, Boland J, et al. Randomized multicenter comparison of conventional anticoagulation versus antiplatelet therapy in unplanned and elective coronary stenting. *Circulation*. 1998:1597−1603. https://doi.org/10.1161/01.cir.98.16.1597.

66. Urban P, Macaya C, Rupprecht HJ, et al. Randomized evaluation of anticoagulation versus antiplatelet therapy after coronary stent implantation in high-risk patients: the multicenter aspirin and ticlopidine trial after intracoronary stenting (MATTIS). *Circulation*. 1998;98(20):2126−2132.

67. Garcia-Garcia HM, McFadden EP, Farb A, et al. Standardized end point definitions for coronary intervention trials: the academic research consortium-2 consensus document. *Circulation*. 2018:2635−2650. https://doi.org/10.1161/circulationaha.117.029289.

68. Yoshida R, Ishii H, Morishima I, et al. Performance of HAS-BLED, ORBIT, PRECISE-DAPT, and PARIS risk score for predicting long-term bleeding events in patients taking

an oral anticoagulant undergoing percutaneous coronary intervention. *J Cardiol*. 2019: 479−487. https://doi.org/10.1016/j.jjcc.2018.10.013.

69. Hulme OL, Khurshid S, Weng L-C, et al. Development and validation of a prediction model for atrial fibrillation using electronic health records. *JACC Clin Electrophysiol*. 2019;5(11):1331−1341.

70. Gibson WJ, Nafee T, Travis R, et al. Machine learning versus traditional risk stratification methods in acute coronary syndrome: a pooled randomized clinical trial analysis. *J Thromb Thrombolysis*. 2020;49(1):1−9.

71. Aradi D, Kirtane A, Bonello L, et al. Bleeding and stent thrombosis on P2Y12-inhibitors: collaborative analysis on the role of platelet reactivity for risk stratification after percutaneous coronary intervention. *Eur Heart J*. 2015;36(27):1762−1771.

72. Price MJ, Endemann S, Gollapudi RR, et al. Prognostic significance of post-clopidogrel platelet reactivity assessed by a point-of-care assay on thrombotic events after drug-eluting stent implantation. *Eur Heart J*. 2008;29(8):992−1000.

73. Price MJ, Berger PB, Teirstein PS, et al. Standard- vs high-dose clopidogrel based on platelet function testing after percutaneous coronary intervention: the GRAVITAS randomized trial. *J Am Med Assoc*. 2011;305(11):1097−1105.

74. Cavallari LH, Lee CR, Beitelshees AL, et al. Multisite investigation of outcomes with implementation of CYP2C19 genotype-guided antiplatelet therapy after percutaneous coronary intervention. *JACC Cardiovasc Interv*. 2018;11(2):181−191.

75. Mega JL, Simon T, Collet J-P, et al. Reduced-function CYP2C19 genotype and risk of adverse clinical outcomes among patients treated with clopidogrel predominantly for PCI: a meta-analysis. *J Am Med Assoc*. 2010;304(16):1821−1830.

76. Claassens DMF, Vos GJA, Bergmeijer TO, et al. A genotype-guided strategy for oral P2Y inhibitors in primary PCI. *N Engl J Med*. 2019;381(17):1621−1631.

77. Sibbing D, Aradi D, Alexopoulos D, et al. Updated expert consensus statement on platelet function and genetic testing for guiding P2Y12 receptor inhibitor treatment in percutaneous coronary intervention. *JACC Cardiovasc Interv*. 2019:1521−1537. https://doi.org/10.1016/j.jcin.2019.03.034.

78. Watanabe H, Domei T, Morimoto T, et al. Effect of 1-month dual antiplatelet therapy followed by clopidogrel vs 12-month dual antiplatelet therapy on cardiovascular and bleeding events in patients receiving PCI: the STOPDAPT-2 randomized clinical trial. *J Am Med Assoc*. 2019;321(24):2414−2427.

79. Franzone A, McFadden E, Leonardi S, et al. Ticagrelor alone versus dual antiplatelet therapy from 1 Month after drug-eluting coronary stenting. *J Am Coll Cardiol*. 2019; 74(18):2223−2234.

Moving from dual antiplatelet therapy to monotherapy based on P2Y$_{12}$ receptor blockade—why it could be a novel paradigm?

Johny Nicolas, MD, Ridhima Goel, MD, Bimmer Claessen, MD, PhD, Roxana Mehran, MD

Center for Interventional Cardiovascular Research and Clinical Trials, Zena and Michael A. Wiener Cardiovascular Institute, Icahn School of Medicine at Mount Sinai, New York, NY, United States

Introduction

Dual antiplatelet therapy (DAPT), consisting of aspirin and a P2Y$_{12}$ receptor inhibitor, is considered mandatory for all patients following percutaneous revascularization or after an acute coronary syndrome (ACS) irrespective of treatment modality.[1,2] Both American and European guidelines recommend initiation of DAPT periprocedurally and its continuation for at least 6 months in stable coronary artery disease (SCAD) and 12 months in ACS.[3] Specific recommendations have been made for two groups of patients: (1) those at high ischemic risk may benefit from prolonged DAPT and (2) those at high bleeding risk (HBR) may benefit from shortened DAPT.[4] Nonetheless, the duration and choice of DAPT are still largely debatable given the advancements made over the past years in medical therapy (i.e., antihypertensive, antianginal, and antidyslipidemic medications), drug-eluting stents (DESs), and antiplatelet agents as compared to the era when guidelines were developed. The rationale behind using a dual antiplatelet regimen stems from the need to provide effective antithrombotic protection, as early discontinuation of DAPT has been associated with an increased risk of stent thrombosis and graft occlusion.[5-7] However, the ischemic protection comes at the expense of an increase in bleeding events that offsets the long-term success of any invasive or noninvasive treatment strategy.[8] Concerns surrounding bleeding, especially in patients with HBR, led to the development of alternative strategies consisting of a shortened DAPT duration followed by single antiplatelet therapy (SAPT). The choice of SAPT is mainly guided by tolerance of antiplatelet agents (i.e., resistance, allergies, side effects, etc.) and could be either aspirin or a $P2Y_{12}$ inhibitor-based regimen. Aspirin monotherapy after a short period of DAPT has been the main strategy for a long time; however, $P2Y_{12}$ inhibitor monotherapy is still a relatively new approach that emerged with the discovery of newer and more potent agents. Recent studies have shown that omitting aspirin from triple antithrombotic therapy (DAPT in addition to an anticoagulant) in patients with atrial fibrillation who are undergoing PCI resulted in less bleeding events as compared to triple therapy, without compromising the antithrombotic protection.[9–11] Consequently, dual antithrombotic therapy without aspirin is now suggested as the optimal regimen after PCI in patients with atrial fibrillation.[12,13] Omission of aspirin after a short period of DAPT in favor of $P2Y_{12}$ inhibitor monotherapy is a novel paradigm that underpins recent and ongoing clinical trials and is the scope of this chapter.

Overview of platelet function and pathophysiology

Platelets play an essential role in primary hemostasis and thrombus formation (see Chapter 1). Under regular physiologic conditions, they circulate in blood in an inactive state. Once endothelial cell injury occurs, platelets are activated and bind via the glycoprotein Ib (GpIb) receptor to von Willebrand factor expressed on vessels' subendothelial collagen. Subsequently, a sequence of responses takes place: change in platelets' shape, secretion of adenosine triphosphate/adenosine diphosphate (ADP) by dense granules, and release of chemokines and coagulation factors by α-granules. The ADP binding to $P2Y_{12}$ receptors amplifies platelets' response and the released chemokines activate endothelial cells and leukocytes. Moreover, the interposition of fibrinogen on the platelets' glycoprotein IIb/IIIa (GpIIb/IIIa) surface receptors induces platelet aggregation.[14,15] This sequence of reactions results in platelet plug formation and maintenance of primary hemostasis.

In addition, platelets participate in the formation and extension of atherosclerotic plaques by adhering to blood vessels' injured endothelium at sites of endothelial cell activation. In response to plaque rupture, platelets mediate the acute onset of arterial thrombosis.[16] Although these physiologic responses are essential to repair damaged tissue, they might progress uncontrolled into amplifying thrombotic loops that lead to intermittent or permanent blood flow obstruction.[17] Hence, inhibition of platelet function is crucial for the prevention of ischemic events, including but not limited to stent thrombosis and myocardial infarction (MI), after stent implantation.

Rationale for dual antiplatelet therapy

Different antiplatelet agents act on different pathways ultimately leading to thrombus formation. The ability of one agent to block platelet aggregation may be overcome by the release of thrombotic factors from platelets and endothelial cells.[18,19] Consequently, the desired therapeutic effect may not be fully achieved by a single antiplatelet agent. Indeed, studies addressing aspirin resistance in patients with ischemic stroke showed that the pharmacologic effect of aspirin is not constant over time, keeping the patient unprotected from ischemia.[20] Moreover, when the dose of aspirin was increased, platelet response was not uniform in over one-third of individuals and ranged from partial to complete inhibition. Therefore aspirin monotherapy provides insufficient protection for high-risk patients and another agent with a different mechanism of action must be added on top of aspirin.

Soon after the implantation of the first coronary stents, postprocedural stent thrombosis emerged as a major complication of the intervention.[21] During that period, no guidelines for optimal antithrombotic therapy existed and warfarin-based anticoagulation after stenting had been used based on experience with prosthetic heart valves. Faced with all these concerns, Colombo and his team[22] succeeded in showing that optimal stent expansion is the single most important factor that directly affects the risk of stent thrombosis regardless of anticoagulation therapy use. Indeed, their work highlighted the safety of withholding anticoagulant use following successful PCI. Moreover, they showed that a dual antiplatelet regimen consisting of ticlopidine and aspirin was superior to aspirin monotherapy in terms of decreasing the incidence of stent thrombosis at 1 month after PCI (0.8% in DAPT group vs. 2.9% in aspirin-only group).[23] Later on, the Stent Anticoagulation Restenosis Study (STARS) confirmed the previous results by showing that the incidence of major adverse cardiac events was lowest among patients treated with aspirin plus ticlopidine (0.5%) as compared with those treated with aspirin and warfarin (2.7%) or aspirin alone (3.6%).[24] Furthermore, the Intracoronary Stenting and Antithrombotic Regimen (ISAR) study established the superiority of DAPT over anticoagulant agents in patients who underwent PCI with stent implantation.[25] Afterward, the CURE (Clopidogrel in Unstable Angina to Prevent Recurrent Events) trial showed that the addition of clopidogrel to aspirin in patients with ACS and in those who underwent PCI was superior to aspirin monotherapy in terms of preventing death, nonfatal MI, or stroke, despite an increase in the risk of bleeding.[26] Following this major trial, the administration of $P2Y_{12}$ inhibitors on top of aspirin became a standard practice and was shown to be associated with favorable outcomes in several studies, although at the expense of greater bleeding.[17,27,28] Nonetheless, the duration of DAPT following PCI has remained a controversial and equivocal topic over the past decades, especially after pooled long-term follow-up data from initial DES approval trials raised concerns about the safety and efficacy of these devices as compared to bare metal stent (BMS). Two meta-analyses presented at the annual meeting of the European Society of Cardiology (ESC) in 2006 showed that the use of first-generation DESs was associated with significantly higher rates of MI and death at 3 years due to an increased incidence of late stent thrombosis as compared with BMS. These controversial results sparked intense discussions among the cardiovascular community and led to the so-called 'ESC firestorm'. Subsequently, guideline recommendations for optimal DAPT duration changed from 3 months (in case of sirolimus-eluting stent) and 6 months (in case of paclitaxel-eluting stent) to 12 months.[29] Yet, with the introduction of newer-generation

DESs with biocompatible polymers, the need for a prolonged DAPT duration following PCI became questionable especially in light of increased recognition of bleeding as a major prognostic factor following PCI.[30,31]

Several recent trials investigated the possibility of shortening the DAPT duration to 6 months[32-37] or 3 months,[38,39] resulting in less bleeding events and similar rates of ischemic events. In contrast, other studies examining the prolongation of DAPT beyond 12 months showed that this strategy results in a lower rate of nonfatal ischemic events but more bleeding-related outcomes.[40-43] A meta-analysis pooling the results of 32,287 patients from 10 randomized trials showed that as compared to a standard 12-month regimen, a shortened DAPT duration (<12 months) reduces major bleeding (odds ratio 0.58; 95% confidence interval [CI], 0.36-0.92; $P = .02$) without compromising the protection from ischemia.[44] Nonetheless, in patients with low bleeding risk but high ischemic risk, a prolonged (>12 months) DAPT duration could be considered, as it significantly reduces the odds of MI (0.53, 95% CI [0.42-0.66], $P < .001$) and stent thrombosis (0.33, 95% CI [0.21-0.51], $P < .001$).

The term 'short DAPT' referred for years to an early discontinuation of the $P2Y_{12}$ inhibitor and continuation of aspirin monotherapy. Nonetheless, the concept of dual therapy can still hold if aspirin is the first agent to be dropped and $P2Y_{12}$ inhibitor is continued as monotherapy. Several reasons favoring the discontinuation of aspirin instead of $P2Y_{12}$ inhibitor currently exist in the literature. As example, gastrointestinal (GI) bleeding and intracranial hemorrhage are the commonly described side effects of aspirin use.[45,46] Yet, several trials and meta-analyses on the use of aspirin for secondary prevention showed that the ischemic benefits outweigh the bleeding risks.[47,48] Of note, all these studies were completed in an era when pharmacologic strategies for secondary prevention were not widely available and implemented. Moreover, potent $P2Y_{12}$ inhibitors such as ticagrelor and prasugrel were not widely used. Consequently, reevaluation of the role of aspirin as the cornerstone of DAPT is much needed.

The rise and fall of aspirin

Salicylate-containing plants were used for their pain-relieving properties by one of the earliest known civilizations in the historical region of southern Mesopotamia around 4000 years ago.[49] Thereafter, salicylates' indications were extended to inflammation and fever. It was not until 1897 that the modern formulation of acetylsalicylic acid was synthesized by the German chemist Felix Hoffmann by adding an acetyl group to salicylic acid.[50] After the discovery of its anti-inflammatory and antithrombotic potentials, aspirin became one of the most prescribed medications.[51] Aspirin exhibits its antithrombotic effect through inhibition of the cyclooxygenase-1 (COX-1) enzyme leading to decreased thromboxane A_2 (TXA_2) synthesis and impaired aggregation of platelets that form the core of the growing thrombotic mass.[52] Indeed, the protective role of aspirin is supported by several studies showing an increased rate of cardiovascular events following drug interruption.[53,54]

The legacy of aspirin

The US Preventive Services Task Force (USPSTF) reports that the benefits of aspirin in preventive medicine are numerous, ranging from colorectal cancer to cardiovascular diseases.[55,56] Indeed, evidence backing aspirin use for secondary prevention of vascular diseases culminated in a 2009 systematic review that pooled data from 16 clinical trials, mostly published before the DES era.[47] The analysis showed that among the 17,000 study participants with a prior history of coronary artery disease (CAD), aspirin use reduced the annual incidence of major vascular events by 18% (6.7% in the aspirin-treated group vs. 8.2% in the control group, $P < .0001$), including coronary events (4.3% vs. 5.3% per year, respectively, $P < .0001$) and strokes (2.1% vs. 2.5% per year, respectively, $P = .002$).[47] Therefore based on high-level evidence available in the literature, clinical guidelines give aspirin a class IA recommendation for secondary prevention in both SCAD and unstable CAD.[4,57−60] Nonetheless, a significantly higher incidence of major bleeding events was observed in aspirin-treated individuals as compared with the control group (0.25% vs. 0.06%, respectively, $P = .01$).[47] Moreover, the high bleeding rates could be even more relevant in elderly people who have a history of GI bleeding and those receiving nonsteroidal anti-inflammatory drugs or other oral anticoagulant agents.

The downfall of aspirin

The adverse effects associated with aspirin use, mainly gastritis and bleeding, are due to COX-1 enzyme inhibition.[61] Indeed, the COX-1 enzyme produces not only TXA_2 responsible for platelet aggregation but also other prostaglandins involved in GI mucosal protection (prostaglandin E_2 and prostaglandin I_2).[61] Due to this deleterious effect, patients maintained on aspirin are at increased risk of GI mucosal erosion and bleeding.[62,63] In addition to gastric toxicity, aspirin has been also strongly associated with an increased incidence of intracranial hemorrhage.[64,65] Moreover, the use of aspirin in primary prevention has been recently challenged by the results of several randomized clinical trials (ARRIVE, ASCEND, and ASPREE).[66−68] A meta-analysis pooling results from 11 trials with 157,248 participants showed that aspirin use in adults without established cardiovascular disease results in a higher incidence of major bleeding without a significant decrease in all-cause mortality.[69] Although the evidence supporting the use of aspirin for secondary prevention is stronger, the increased relevance of bleeding as a major prognostic factor after PCI pushed clinical research toward other safer alternatives such as discontinuation of aspirin at the earliest possible opportunity.[1,27,48,70,71]

Rationale for aspirin withdrawal
Evidence at the molecular level

In theory, aspirin and $P2Y_{12}$ inhibitors impair platelets function by acting on different targets in the platelet activation pathway. Aspirin inhibits the activity of COX-1 enzyme leading to a decreased production of prostaglandins and TXA_2.[72] In contrast, $P2Y_{12}$ inhibitors act on its corresponding platelet receptors preventing the binding of ADP and leading to decreased platelet activation as well as aggregation in response to stimuli.[73] The dual therapy strategy results in an additive or synergistic inhibition of platelet function, as both drugs independently block the TXA_2- and $P2Y_{12}$-dependent pathways.[74–77] However, some studies have shown that $P2Y_{12}$ receptor blockade has an indirect impact on the TXA_2-dependent pathways through inhibition of TXA_2-induced ADP release.[78–80] Moreover, inhibition of $P2Y_{12}$ receptor also affects the TXA_2-independent processes such as thrombin production and thrombin receptor activation.[81,82] Hence, the effect achieved with aspirin use on top of $P2Y_{12}$ antagonists could be minimal rather than synergistic as previously suggested.[83–85] Armstrong et al.[80] showed that the activation of $P2Y_{12}$ receptor in vitro is important to platelet TXA_2 production. Furthermore, in the presence of a potent $P2Y_{12}$ antagonist, only a minimal additional inhibition is achieved with the use of other platelet agonists.[86] All the previously described evidence raised the hypothesis that the use of a single potent $P2Y_{12}$ antagonist without aspirin could be sufficient to decrease platelet reactivity and prevent the occurrence of thrombotic events.[86,87]

Evidence from clinical trials

Aspirin was always the background therapy in the earliest randomized trials evaluating the efficacy of antithrombotic agents as part of a dual or triple therapy regimen. Its efficacy vis-à-vis other antiplatelet agents such as $P2Y_{12}$ antagonists was assessed only in a few studies. In 1989, the TASS (Ticlopidine vs. Aspirin for the prevention of recurrent stroke study) randomized clinical trial showed that ticlopidine was associated with a lower rate of ischemic events and slightly less GI bleeding events as compared with high-dose aspirin (1300 mg) in 3069 patients at high risk of stroke.[88] In 1996, a large randomized clinical trial, CAPRIE (Clopidogrel vs. Aspirin in patients at high risk of ischemic events), showed that clopidogrel use in 19,185 patients with a recent history of MI, stroke, or symptomatic peripheral artery disease resulted in a decrease in both ischemic and GI bleeding events as compared to aspirin.[89] Another study supporting the use of $P2Y_{12}$ inhibitor monotherapy is the MATCH randomized, double-blind, and placebo-controlled trial of 7599 patients that compared aspirin and clopidogrel dual therapy with clopidogrel monotherapy in patients with recent ischemic stroke or transient ischemic attack.[90] Indeed, this study showed that adding aspirin on top of clopidogrel did not result in a significant decrease in cerebrovascular events (relative risk reduction 6.4%, 95% CI

[−4.6 to 16.3]) but increased the incidence of life-threatening bleeding events (absolute risk increase 1.3%, 95% CI [0.6−1.9]).[90] In the Assessment of Dual Anti-Platelet Therapy with Drug-Eluting Stents (ADAPT-DES) prospective registry, a high platelet reactivity on aspirin was not associated with adverse ischemic outcomes such as death, MI, or stent thrombosis (hazard ratio [HR] 1.46; 95% CI, 0.58−3.64; $P = .42$) at 1 year but was inversely related to bleeding (HR 0.65, 95% CI, 0.43−0.99, $P = .04$).[91] Moreover, aspirin resistance was not linked with adverse outcomes in patients on aspirin monotherapy.[92] Hence, the real benefit of aspirin as part of the antiplatelet regimen following PCI in the DES era seems limited.

Discontinuation of DAPT after PCI includes not only stopping both agents but also switching to $P2Y_{12}$ receptor inhibitor or aspirin monotherapy. In the j-Cypher registry of 12,812 patients who have undergone first-generation DES implantation, discontinuation of both aspirin and $P2Y_{12}$ receptor inhibitor resulted in an increase in the incidence of stent thrombosis up to 5 years after PCI.[93] These findings were not observed in patients who were switched to monotherapy at different study time intervals. Similarly, in the CREDO-Kyoto Registry Cohort-2 with 10,470 patients undergoing PCI either with BMS or sirolimus-eluting stent implantation, the incidence of ischemic events (including stent thrombosis, spontaneous MI, and stroke) was higher among patients who stopped both antiplatelet agents (aspirin and $P2Y_{12}$ receptor inhibitor) than only one agent (aspirin or $P2Y_{12}$ receptor inhibitor).[94]

Following these studies, a new set of trials emerged to investigate aspirin-free strategies in patients undergoing PCI (Table 10.1). Most of these randomized trials investigated the feasibility of dropping aspirin and continuing with $P2Y_{12}$ inhibitor monotherapy after a short period of DAPT. The very first trials exploring this strategy (WOEST, PIONEER-AF PCI, RE-DUAL PCI, and AUGUSTUS) included patients with baseline medical conditions requiring oral anticoagulants on top of DAPT, which places them at HBR according to the Academic Research Consortium (ARC)-HBR criteria.[9−11,95] A meta-analysis of these four trials (n = 10,026) showed that aspirin discontinuation soon after PCI resulted in fewer bleeding complications without compromising the ischemic protection as compared to a triple therapy regimen.[96] Extrapolation of aspirin discontinuation to patients on DAPT seemed to be an intriguing concept, especially with the increased awareness of bleeding complications and access to newer and more potent $P2Y_{12}$ inhibitors as well as cutting-edge stent technology. Opponents of this strategy argue that $P2Y_{12}$ inhibitors exhibit a certain degree of interindividual variability in terms of platelet response to antagonists. In particular, patients who are resistant to clopidogrel might not be fully protected after discontinuation of aspirin. Nonetheless, these arguments were more or less addressed in the CAPRIE trial that showed no significant increase in ischemic risk in patients with clopidogrel resistance, but rather a significant reduction in the combined endpoint of ischemic stroke, MI, or vascular death with clopidogrel versus aspirin.[97]

Table 10.1 Randomized clinical trials evaluating P2Y$_{12}$ inhibitor monotherapy in patients undergoing PCI.

	TWILIGHT[107]	GLOBAL LEADERS[105]	GLASSY[106]	STOP DAPT-2[100]	SMART CHOICE[98]	TICO[109]
Key inclusion criteria	Patients undergoing PCI with ≥1 DES and having ≥1 clinical and ≥1 angiographic criteria.	Patients undergoing PCI for coronary stenosis ≥50% with RVD ≥2.25 mm.	Patients undergoing PCI for coronary stenosis ≥50% with RVD ≥2.25 mm.	Patients undergoing PCI with CoCr-EES.	Patients undergoing PCI with DES for ≥50% stenosis with RVD ≥2.25 mm and ≤4.25 mm.	Patients with ACS undergoing PCI with BP-SES.
Key exclusion criteria	STEMI presentation; cardiogenic shock or hemodynamic instability; requirement for DAPT or OAC; history of major bleed, ICH, or bleeding diathesis; planned CABG.	Planned CABG, requirement for DAPT or OAC, history of ICH or major bleed, PCI for prior stent thrombosis.	Planned CABG, requirement for DAPT or OAC, history of ICH or major bleed, PCI for prior stent thrombosis.	Requirement for DAPT or OAC, history of ICH, periprocedural serious complication, DES other than Xience CoCr-EES.	Hemodynamic instability or cardiogenic shock, active gastrointestinal bleeding or genitourinary bleeding, prior DES in 12 months.	Requirement for OAC; history of prior hemorrhagic stroke, brain tumor, aortic dissection, active bleeding, or bleeding diathesis; 1 year history of stroke or CNS impairment; 6-month history of traumatic brain injury; 6-week history of internal bleeding.
Status	Completed	Completed	Competed	Completed 1-year follow-up	Completed 1-year follow-up	Completed
Enrolling years	July 2015 December 2017	July 2013 November 2015	July 2013 November 2015	December 2015 December 2017	March 2014 July 2017	August 2015 October 2019
Enrolling region	187 Sites in 11 countries (Asia, Europe, and North America)	130 Sites in 18 countries (Asia, Australia, Europe, North America, and South America)	20 Highest recruiting sites in the GLOBAL LEADERS study	East Asia: 90 sites in Japan	East Asia: 33 sites in South Korea	East Asia: sites in Korea

Hypothesis	Superiority of 12 months of SAPT with ticagrelor after 3 months of DAPT with ticagrelor and ASA versus 15 months of DAPT with ticagrelor and ASA for reducing 1-year clinically relevant bleeding after PCI.	Superiority of 23 months of SAPT with ticagrelor after 1-month DAPT with ticagrelor and ASA versus standard DAPT regimens for 24 months for preserving ischemic protection after PCI.	Noninferiority, and sequentially, superiority, of 23 months of SAPT with ticagrelor after 1-month DAPT with ticagrelor and ASA versus standard DAPT regimens for 24 months in reducing ischemic events after PCI.	Noninferiority of 11-month SAPT with clopidogrel after 1-month DAPT versus 12 months of DAPT for safety after PCI with CoCr-EES.	Noninferiority of $P2Y_{12}$-I SAPT after 3-month DAPT versus 12-month DAPT in reducing ischemic events and bleeding risk after PCI with DES.	Superiority of 9-month ticagrelor SAPT after 3-month DAPT with ticagrelor and ASA versus 12-month DAPT with ticagrelor and ASA for net adverse outcomes after PCI up to 12 months;
Design	Randomized, placebo-controlled, double-blinded.	Randomized, open-label.	Adjudication substudy of GLOBAL LEADERS.	Randomized, open-label.	Randomized, open-label.	Randomized, open-label.
Randomized patients	Overall: n = 7119 Control: n = 3564 Experimental: n = 3555	Overall: n = 15,968 Control: n = 7988 Experimental: n = 7980	Overall: n = 7585 Control: n = 3791 Experimental: n = 3794	Overall: n = 3045 Control: n = 1552 Experimental: n = 1523	Overall: n = 2993 Control: n = 1498 Experimental: n = 1495	Overall: n = 3056

Continued

Table 10.1 Randomized clinical trials evaluating P2Y$_{12}$ inhibitor monotherapy in patients undergoing PCI.—*cont'd*

	TWILIGHT[107]	GLOBAL LEADERS[105]	GLASSY[106]	STOP DAPT-2[100]	SMART CHOICE[98]	TICO[109]
Control group	Ticagrelor 90 mg bid + ASA 81 –100 mg OD for 15 months	ACS: ticagrelor 90 mg OD for 1 year + ASA 75–100 mg OD for 2 years Or SCAD: clopidogrel 75 mg OD for 1 year + ASA 75–100 mg OD for 2 years	ACS: ticagrelor 90 mg bid for 1 year + ASA 75–100 mg OD for 2 years Or SCAD: clopidogrel 75 mg OD × 1 year + ASA 75–100 mg OD x 2 years	Clopidogrel 75 mg OD or prasugrel 3.75 mg OD for 1 year + ASA 81 –200 mg OD for 5 years	P2Y$_{12}$–I for 12 months + ASA 100 mg OD for 12 months	Ticagrelor 90 mg bid + ASA 100 mg OD for 15 months
Experimental group	Ticagrelor 90 mg bid for 15 months + ASA 81–100 mg OD for 3 months	Ticagrelor 90 mg bid for 2 years + ASA 75–100 mg OD for 1 month	Ticagrelor 90 mg bid for 2 years + ASA 75–100 mg OD for 1 month	Clopidogrel 75 mg OD or prasugrel 3.75 mg OD for 5 years + ASA 81 –200 mg OD for 1 month	P2Y$_{12}$–I for 12 months + ASA 100 mg OD for 3 months	Ticagrelor 90 mg bid for 15 months + ASA 100 mg OD for 3 months
Primary endpoint	BARC type 2, 3, or 5 bleeding up to 1 year	Composite of death or nonfatal new Q-wave MI up to 2 years	Efficacy: composite of death, nonfatal MI, stroke, or urgent TVR at 2 years Safety: BARC 3 or 5 bleeding at 2 years	Composite of CVD, MI, stroke, ST, or TIMI major/ minor bleeding up to 1 year	Composite of death, MI, or CVA after PCI up to 1 year	Composite of death, MI, ST, CVA, TVR, or TIMI major bleeding up to 12 months after PCI
Key secondary endpoint	Composite of death from any cause, nonfatal MI or CVA up to 1 year	BARC 3 or 5 bleeding up to 2 years	–	Composite CVD, MI, stroke, ST; TIMI major/Minor bleeding up to 1 year and 5 years	BARC 2–5 bleeding up to 1 year	Components of 1° Endpoint

Follow-up periods	3, 4, 9, 15, and 18 months	1, 3, 6, 12, 18, and 24 months	1, 3, 6, 12, 18, and 24 months	1, 12, 24, 36, 48, and 60 months	3, 6 12, 24, and 36 months	3, 6, 9, and 12 months
Results	**Primary endpoint:** 7.1% (control) and 4.0% (experimental); HR = 0.56 (0.45 −0.68), $P < .001$ **Secondary endpoint:** 3.9% (DAPT) and 3.9% (SAPT); HR = 0.99 (0.78−1.25)	**Primary endpoint:** 4.37% (control) and 3.81% (experimental); RR = 0.87 (0.75 −1.01), $P = .073$ **Secondary endpoint:** 2.04% (exp) and 2.12% (ref); RR = 0.97 (0.78 −1.20), $P = .77$	**Primary efficacy endpoint:** 8.41% (control) and 7.14% (experimental); RR = 0.848 (0.721−0.998), $P = .047$ **Primary safety endpoint:** 2.46% (control) and 2.46% (experimental)	**Primary endpoint:** 3.70% (control) and 2.36% (experimental); HR = 0.64 (0.42 −0.98) **Secondary ischemic endpoint:** 2.51% (control) and 1.96% (experimental); HR = 0.79 (0.49 −1.29) **Secondary bleeding endpoint:** 1.54% (control) and 0.41% (experimental); HR = 0.26 (0.11 −0.64)	**Primary endpoint:** 2.5% (control) and 2.9% (experimental); HR = 1.19 (0.76 −1.85) **Secondary bleeding endpoint:** 3.4% (control) and 2.0% (experimental); HR = 0.58 (0.36 −0.92)	—

ACS, acute coronary syndrome; ASA, acetylsalicylic acid; BARC, Bleeding Academic Research Consortium; bid, twice daily; BP-SES, bioresorbable polymer sirolimus-eluting stent; CABG, coronary artery bypass graft surgery; CNS, central nervous system; CoCr-EES, cobalt chromium everolimus-eluting stent; CVA, cerebrovascular accident; CVD, cardiovascular disease; DAPT, dual antiplatelet therapy; DES, drug-eluting stent; HR, hazard ratio; ICH, intracranial hemorrhage; MI, myocardial infarction; OAC, oral anticoagulation; OD, once daily; $P2Y_{12}$-I, $P2Y_{12}$ inhibitor; PCI, percutaneous coronary intervention; RR, relative risk; RVD, reference vessel diameter; SAPT, single antiplatelet therapy; SCAD, stable coronary artery disease; ST, stent thrombosis; STEMI, ST-segment elevation myocardial infarction; TIMI, Thrombolysis in Myocardial Infarction; TVR, target vessel revascularization.

Clinical trials evaluating P2Y$_{12}$ inhibitor monotherapy

Based on the theoretic construct described earlier, several clinical trials (Table 10.1) have been conducted to reevaluate the role of aspirin in secondary prevention in the presence of potent P2Y$_{12}$ inhibitors and cutting-edge stent technology.

SMART-CHOICE

The open-label SMART-CHOICE trial (Comparison Between P2Y12 Antagonist Monotherapy and Dual Antiplatelet Therapy After DES, NCT02079194) randomized 2993 South Korean patients who underwent PCI with current-generation DESs (Xience, Promus, Synergy, and Orsiro) to either 3 (n = 1495, experimental arm) or 12 months (n = 1498, control arm) of DAPT.[98] At 90 days after randomization, aspirin was stopped in the first group and patients were maintained on P2Y$_{12}$ inhibitor monotherapy (76.9% on clopidogrel, 19.0% on ticagrelor, and 4.1% on prasugrel). The primary outcome of interest, major adverse cardiac and cerebrovascular events, occurred in 2.9% of participants in the experimental group and 2.5% of participants in the 12-month DAPT group (P for no inferiority = .007 and P for superiority = .28). The results were confirmed in a landmark analysis at 90 days. The secondary endpoints (all-cause death, MI, stroke, cardiac death, and stent thrombosis) were in concordance between the two groups at 12 months, with the exception of a higher incidence of Bleeding Academic Research Consortium (BARC) types 2–5 bleeding observed in the prolonged DAPT group (2.0% vs. 3.4%; HR 0.36, 95% CI [0.34–1.01], P = .02). The results were consistent among all prespecified subgroups (age, sex, ACS, chronic renal failure, diabetes, previous stroke, low left ventricular ejection, multivessel PCI, P2Y$_{12}$ inhibitor, and type of DES). This trial showed that P2Y$_{12}$ inhibitor monotherapy for 12 months after 3 months of DAPT is a novel antithrombotic strategy that balances the ischemic and bleeding risks in patients who received a DES. Nonetheless, this study had some limitations, as around 10% of patients randomized to monotherapy were still taking aspirin at 12 months. Moreover, this trial included all-comer patients, and although no statistical interaction was observed between patients with ACS versus those without ACS and randomized treatment assignment in terms of the primary outcome, the trial is inherently underpowered to show noninferiority in the ACS subgroup who may theoretically benefit from a prolonged DAPT duration.[99]

STOPDAPT-2

The open-label STOPDAPT-2 (Short and optimal duration of dual antiplatelet therapy-2 study, NCT02619760) randomized trial enrolled 3045 patients following PCI with a cobalt-chromium everolimus eluting stent (EES, Xience, Abbott) at 90 hospitals across Japan.[100] Patients were randomized in a 1:1 fashion after PCI to receive DAPT consisting of clopidogrel or prasugrel on top of aspirin for 1 versus 12 months. In the 1-month DAPT group, patients were to receive clopidogrel monotherapy up to 5 years after discontinuation of aspirin. In the 12-month DAPT group, patients were to stop clopidogrel after 12 months and receive aspirin monotherapy for up to 5 years. The primary outcome of interest, i.e., net adverse cardiovascular events, which is a composite of cardiovascular death, MI, stent thrombosis, stroke, or Thrombolysis in Myocardial Infarction (TIMI) major/minor bleeding, occurred more frequently in patients on dual therapy at 1-year follow-up (3.7% in 12-month DAPT group vs. 2.4% in 1-month DAPT group, P for superiority = .04). In addition, patients on prolonged DAPT had more TIMI major/minor bleeding events (1.5% vs. 0.4%, P for superiority = .004). Both groups had similar rates of adverse events at 1 year with a very low overall stent thrombosis rate (1-month DAPT 0.27% vs. 12-month DAPT 0.07%, $P = .21$). The results were consistent across the different subgroups analyzed with the exception of patients with chronic kidney disease who did better on a prolonged DAPT regimen. In summary, 1 month of DAPT followed by clopidogrel monotherapy resulted in a significantly lower rate of a composite of ischemic and bleeding events at 1 year as compared with a prolonged dual therapy. Of note, around 97% of all treated patients underwent intravascular imaging (intravascular ultrasound or optical coherence tomography) during the procedure. Consequently, the low event rates may be in part due to this common practice in Japan, which could affect the generalizability of the results to patients in Europe and North America where these imaging techniques are used in a minority of procedures.

Indeed, both the SMART-CHOICE and STOPDAPT-2 randomized trials had some limitations that prevent extrapolation of the results to patients treated outside Asia. All participants enrolled in these two studies are East Asians who are known to have particular genetic features that directly affect the pharmacotherapeutic response to different antiplatelet agents.[101,102] For example, a high prevalence of CYP2C19 genetic polymorphism is seen among East Asian populations leading to poor metabolism of clopidogrel.[103]

GLOBAL LEADERS

The GLOBAL LEADERS (GLOBAL LEADERS: A Clinical Study Comparing Two Forms of Anti-platelet Therapy After Stent Implantation; NCT01813435) was an open-label, superiority trial conducted at 130 sites across 18 countries in Pacific Asia, Europe, North America, and South America.[104] The study recruited all-comer patients who underwent PCI with a biolimus-eluting stent for SCAD or ACSs.[105] Participants (n = 15,968) were randomly assigned (1:1 ratio) to the control arm (n = 7767), where they received DAPT with aspirin plus clopidogrel (in patients with SCAD) or ticagrelor (in patients with ACS) for 12 months followed by aspirin monotherapy for an additional 12-month period, or to the experimental arm (n = 7782), where they received DAPT with aspirin plus ticagrelor for 1 month followed by ticagrelor monotherapy for 23 months, essentially comparing aspirin and ticagrelor monotherapy for the second year of the trial. The primary endpoint of interest (all-cause mortality or nonfatal MI) occurred in 4.4% of the control group and 3.8% of the experimental group ($P = .073$). Bleeding grade 3 or 5 (according to the BARC) occurred similarly in both groups (2.1% vs. 2.0%, respectively, 95% CI [0.8−1.2], $P = .77$). The GLASSY (GLOBAL LEADERS Adjudication Sub-StudY, NCT03231059) is a subanalysis of the main GLOBAL LEADERS trial in which investigator-reported outcomes were centrally adjudicated by a clinical event committee.[106] This study showed that the primary efficacy endpoint (all-cause death, MI, stroke, or urgent target vessel revascularization) occurred in 8.4% of the control group and 7.1% of the experimental group (P-noninferiority $<.001$ and P-superiority $= .047$). Similarly, BARC 3 or 5 bleeding occurred at similar rates in both groups (2.5% vs. 2.5%, respectively, $P = .99$). A landmark analysis was performed at 1 year and showed a benefit favoring the experimental arm for the prevention of very late stent thrombosis (P-interaction $= .007$).

TWILIGHT

The TWILIGHT (Ticagrelor with aspirin or alone in high-risk patients after coronary intervention; NCT02270242) study was a randomized, double-blind clinical trial conducted in 187 sites across 11 countries.[107] Study participants received ticagrelor (90 mg twice daily) and aspirin (81−100 mg daily) for the first 3 months after index PCI. Those who were event-free and adherent at 3 months were randomized to either ticagrelor on top of aspirin or ticagrelor monotherapy for an additional 12 months. The primary endpoint of interest, BARC type 2, 3, or 5 bleeding, occurred less often in the experimental arm than in the control arm (4.0% vs. 7.1%, respectively, HR 0.56; 95% CI [0.45−0.68]).[107] Notably, the combined endpoint of death, MI, or stroke occurred similarly in both arms (3.9% vs. 3.9%, respectively, HR 0.99; 95% CI [0.78−1.25], P for noninferiority is $<.001$).[107] In addition, prespecified subgroup analyses showed consistent benefits across different variables including sex, age, diabetes mellitus, total stent length, number of diseased vessels, and geographic region. Hence, TWILIGHT showed that in high-risk patients undergoing PCI, a short DAPT duration of 3 months followed by ticagrelor monotherapy resulted in less bleeding with similar rates of ischemic events as compared to a dual therapy strategy. Nonetheless, TWILIGHT's results cannot be generalized to all patients undergoing PCI because the study had very specific patient inclusion and exclusion criteria, as listed in Table 10.2. Moreover, the results cannot be generalized to other P2Y$_{12}$ inhibitors such as clopidogrel and prasugrel.

Concurrent with the main trial, a thrombogenicity substudy was conducted to provide mechanistic and physiologic basis for the tested hypothesis. From all patients enrolled in TWILIGHT, 51 participants underwent additional blood thrombogenicity and platelet reactivity testing at random times, i.e., 4 and 9 months after the enrollment date.[108] The primary endpoints of interest were platelet-dependent thrombus area measured by the Badimon perfusion chamber (a model that generates thrombus under dynamic flow conditions of shear stress similar to CAD) and platelet reactivity measured by the Multiplate Analyzer (DiaPharma, West Chester, Ohio). For the first endpoint, there was no significant difference in the ex vivo thrombus

Table 10.2 TWILIGHT trial inclusion criteria.

High-risk criteria (Patient should have at least one clinical and one angiographic criteria to be considered at high risk)	
Clinical criteria	**Angiographic criteria**
Age at least 65 years	Multivessel coronary artery disease
Female sex	Total stent length more than 30 mm
Troponin-positive acute coronary syndrome	Thrombotic target lesion
Established vascular disease	Bifurcation lesion treated with two stents
Diabetes mellitus treated with medications	Obstructive left main or proximal left anterior descending lesion
Chronic kidney disease	Calcified target lesion treated with atherectomy

Adapted from Mehran R, Baber U, Sharma SK et al. Ticagrelor with or without aspirin in high-risk patients after PCI. N Engl J Med. 2019;381(21):2032–2042.

area between the two groups (mean difference of 218.2 μm^2; 95% CI [−575.9 to 139.9], $P = .22$).[108] Similarly, no significant difference was noted in platelet reactivity using either the thrombin receptor activator peptide 6 ($P = .81$) or ADP ($P = .47$).[108] However, the experimental arm had higher platelet aggregation when the arachidonic acid ($P = .02$) and collagen ($P = .03$) agonists were used. The mechanistic substudy confirmed the main trial results, as it was shown that aspirin discontinuation does not result in an increase in thrombus formation leading to a greater risk of ischemic events.

TICO

The TICO study (Ticagrelor after 3 Months in the Patients Treated with New Generation Sirolimus Stent for Acute Coronary Syndrome - NCT 02494895) was an open-label randomized trial of 3056 patients with ACS who underwent PCI with bioresorbable sirolimus-eluting stent. All participants received DAPT consisting of aspirin and ticagrelor for 3 months and were then randomized to either ticagrelor monotherapy or continued DAPT for another 9 months. The primary outcome was NACE (Net Adverse Clinical Events), a combination of major bleeding and MACE at 12 months, with prespecified landmark analyses after 3 months of DAPT. The primary endpoint occurred in 3.9% of monotherapy and 5.9% of DAPT patients (HR 0.66, 95% CI [0.48−0.92], $P = .01$). In a landmark analysis at 3 months, the respective rates were 1.4% and 3.5% (HR 0.41, 95% CI [0.25−0.68], $P = .001$). There was a substantial reduction in major bleeding (1.7% vs. 3.0%, HR 0.56, 95% CI [0.34−0.91], $P = .02$, respectively), with all the difference occurring after 3 months (0.2% vs. 1.6%, HR 0.13, 95% CI [0.04−0.44], $P = .001$, respectively). There were no significant differences in MACE, although monotherapy patients fared numerically better in this category too.

Ongoing and future P2Y$_{12}$ inhibitor monotherapy trials

The ASET (Acetyl Salicylic Elimination Trial, NCT033469856) proof-of-concept pilot study evaluating the use of prasugrel monotherapy following new-generation biodegradable polymer DES (SYNERGY stent) implantation in patients with SCAD or stabilized ACS.[110] In this open-label, single-arm study, eligible patients are loaded with a standard DAPT regimen consisting of aspirin and clopidogrel 2 h prior to PCI. Whenever the procedure is deemed successful based on specific clinical and angiographic findings, patients are placed on prasugrel monotherapy for 3 months after which they will switch to either DAPT or aspirin monotherapy at the discretion of their treating physician. The primary endpoints of interest are ischemic (composite of cardiac death, target vessel MI, and stent thrombosis) and bleeding (BARC types 3 and 5) outcomes at 3 months. In fact, ASET is the first trial to examine the efficacy and safety of completely removing aspirin from the antiplatelet regimen immediately after PCI.

Given the aging population presenting for PCI and the high prevalence of comorbidities among them, an antiplatelet regimen that balances the risk of postprocedural ischemic and bleeding complications is imperative. Evidence from recent clinical trials shows that after a short period of DAPT, P2Y$_{12}$ inhibitor monotherapy is as efficacious as a dual regimen including aspirin in maintaining this balance. Owing to the increased awareness of bleeding as a major cause of morbidity and mortality following PCI, a new consensus statement has been recently released by the ARC to identify patients considered at HBR (ARC-HBR) periprocedurally and following PCI.[111] Indeed, the ARC-HBR consortium lists a set of major and minor criteria (Table 10.3) that put a patient at high risk of major bleeding. The purpose of this

Table 10.3 ARC-HBR consensus major and minor criteria.

ARC-HBR criteria (at least one major and two minor criteria are met)	
Major HBR criteria	**Minor HBR criteria**
- Active malignancy (excluding nonmelanoma skin cancer) diagnosed within the past 12 months	- Moderate CKD (eGFR 30—59 mL/min)
- Advanced CKD (eGFR <30 mL/min)	- Hemoglobin: 11—12.9 g/dL for men and 11—11.9 g/dL for women
- Hemoglobin <11 g/dL	- Age ≥75 years
- Liver cirrhosis with portal hypertension	- Long-term use of oral NSAIDs or steroids
- Long-term oral anticoagulation (not including vascular protection doses)	- Spontaneous bleeding requiring hospitalization or transfusion within the past 12 months not meeting the major criterion
- Moderate/severe baseline thrombocytopenia (<100 × 10^9/L)	- Any ischemic stroke at any time not meeting the major criterion

Table 10.3 ARC-HBR consensus major and minor criteria.—*cont'd*

ARC-HBR criteria (at least one major and two minor criteria are met)	
Major HBR criteria	**Minor HBR criteria**
- Nondeferrable major surgery on DAPT - Previous spontaneous ICH/traumatic ICH (within the past 12 months)/ presence of a bAVM/moderate or severe ischemic stroke within the past 6 months - Recent major surgery or major trauma within 30 days before PCI - Spontaneous bleeding requiring hospitalization or transfusion in the past 6 months or at any time (if recurrent) - Tendency to bleed or bruise easily	

ARC-HBR, *Academic Research Consortium-high bleeding risk;* bAVM, *brain arteriovenous malformation;* CKD, *chronic kidney disease;* DAPT, *dual antiplatelet therapy;* eGFR, *estimated glomerular filtration rate;* ICH, *intracranial hemorrhage;* NSAID, *nonsteroidal anti-inflammatory drugs;* PCI, *percutaneous coronary intervention.*

Adapted from Urban P, Mehran R, Colleran R et al. Defining high bleeding risk in patients undergoing percutaneous coronary intervention: a consensus document from the Academic Research Consortium for High Bleeding Risk. Eur Heart J 2019;40(31):2632—2653.

document is to standardize the inclusion and exclusion criteria among all future trials, which helps in the comparison and interpretation of study results. These trials are expected to provide clinicians with a better understanding of the efficacy and safety of $P2Y_{12}$ inhibitor monotherapy in a vulnerable group of patients who are at increased risk of various postprocedural complications.

Conclusion

Aspirin is a widely accessible and affordable medication and its role in secondary prevention has been well established in major trials. Yet, in the era of highly refined revascularization techniques and potent $P2Y_{12}$ inhibitors, aspirin's added net value becomes uncertain especially in light of its adverse effects. Indeed, an aspirin-free strategy after a short period of DAPT seems a reasonable strategy, especially in patients with HBR. $P2Y_{12}$ inhibitor monotherapy is emerging as a plausible approach and is currently under investigation in several trials. Nonetheless, the optimal DAPT duration and the choice of $P2Y_{12}$ inhibitor agent remain unknown and tend to be highly dependent on each patient's clinical and procedural features. Further research is needed to address these issues and establish new guidelines for the optimal antiplatelet therapy following PCI and ACS. The recently published ARC-HBR consensus document will guide the design of the next clinical trials examining the safety and efficacy of $P2Y_{12}$ inhibitor monotherapy in these vulnerable populations.

Finally, future studies should address the cost-effectiveness of an aspirin-free strategy in favor of a potent P2Y$_{12}$ inhibitor, as aspirin remains a largely accessible medication and is available at a lower cost than P2Y$_{12}$ inhibitors.

References

1. Levine GN, Bates ER, Bittl JA, et al. 2016 ACC/AHA guideline focused update on duration of dual antiplatelet therapy in patients with coronary artery disease: a report of the American College of Cardiology/American Heart Association Task Force on Clinical Practice Guidelines. *Circulation*. 2016;68(10):1082−1115.
2. Valgimigli M, Bueno H, Byrne RA, et al. 2017 ESC focused update on dual antiplatelet therapy in coronary artery disease developed in collaboration with EACTS. *Kardiol Pol*. 2017;75(12):1217−1299.
3. Capodanno D, Alfonso F, Levine GN, Valgimigli M, Angiolillo DJ. ACC/AHA versus ESC guidelines on dual antiplatelet therapy: JACC guideline comparison. *J Am Coll Cardiol*. 2018;72(23 Pt A):2915−2931.
4. Levine GN, Bates ER, Bittl JA, et al. 2016 ACC/AHA guideline focused update on duration of dual antiplatelet therapy in patients with coronary artery disease: a report of the American College of Cardiology/American Heart Association Task Force on Clinical Practice Guidelines: an update of the 2011 ACCF/AHA/SCAI guideline for percutaneous coronary intervention, 2011 ACCF/AHA guideline for coronary artery bypass graft surgery, 2012 ACC/AHA/ACP/AATS/PCNA/SCAI/STS guideline for the diagnosis and management of patients with stable ischemic heart disease, 2013 ACCF/AHA guideline for the management of ST-elevation myocardial infarction, 2014 AHA/ACC guideline for the management of patients with non−ST-elevation acute coronary syndromes, and 2014 ACC/AHA guideline on perioperative cardiovascular evaluation and management of patients undergoing noncardiac surgery. *Circulation*. 2016; 134(10):e123−e155.
5. Mehran R, Baber U, Steg PG, et al. Cessation of dual antiplatelet treatment and cardiac events after percutaneous coronary intervention (PARIS): 2 year results from a prospective observational study. *Lancet*. 2013;382(9906):1714−1722.
6. van Werkum JW, Heestermans AA, Zomer AC, et al. Predictors of coronary stent thrombosis: the Dutch stent thrombosis registry. *J Am Coll Cardiol*. 2009;53(16):1399−1409.
7. Saw J, Wong GC, Mayo J, et al. Ticagrelor and aspirin for the prevention of cardiovascular events after coronary artery bypass graft surgery. *Heart*. 2016;102(10): 763−769.
8. Mehran R, Pocock SJ, Stone GW, et al. Associations of major bleeding and myocardial infarction with the incidence and timing of mortality in patients presenting with non-ST-elevation acute coronary syndromes: a risk model from the ACUITY trial. *Eur Heart J*. 2009;30(12):1457−1466.
9. Dewilde WJM, Oirbans T, Verheugt FWA, et al. Use of clopidogrel with or without aspirin in patients taking oral anticoagulant therapy and undergoing percutaneous coronary intervention: an open-label, randomised, controlled trial. *Lancet*. 2013; 381(9872):1107−1115.
10. Gibson CM, Mehran R, Bode C, et al. Prevention of bleeding in patients with atrial fibrillation undergoing PCI. *N Engl J Med*. 2016;375(25):2423−2434.

11. Cannon CP, Bhatt DL, Oldgren J, et al. Dual antithrombotic therapy with dabigatran after PCI in atrial fibrillation. *N Engl J Med*. 2017;377(16):1513−1524.

12. Angiolillo DJ, Goodman SG, Bhatt DL, et al. Antithrombotic therapy in patients with atrial fibrillation undergoing percutaneous coronary intervention: a North American perspective—2016 update. *Circulation*. 2016;9(11):e004395.

13. January CT, Wann LS, Calkins H, et al. 2019 AHA/ACC/HRS focused update of the 2014 AHA/ACC/HRS guideline for the management of patients with atrial fibrillation: a report of the American College of Cardiology/American Heart Association Task Force on Clinical Practice Guidelines and the Heart Rhythm Society. *J Am Coll Cardiol*. 2019; 74(1):104−132.

14. Angiolillo DJ, Ueno M, Goto S. Basic principles of platelet biology and clinical implications. *Circ J*. 2010;74(4):597−607.

15. Angiolillo DJ, Capodanno D, Goto S. Platelet thrombin receptor antagonism and atherothrombosis. *Eur Heart J*. 2010;31(1):17−28.

16. Davi G, Patrono C. Mechanisms of disease: platelet activation and atherothrombosis. *N Engl J Med*. 2007;357(24):2482−2494.

17. Patrono C, Andreotti F, Arnesen H, et al. Antiplatelet agents for the treatment and prevention of atherothrombosis. *Eur Heart J*. 2011;32(23):2922−2932.

18. Libby P. Multiple mechanisms of thrombosis complicating atherosclerotic plaques. *Clin Cardiol*. 2000;23(Suppl 6(S6)). VI-3-7.

19. Altman R, Scazziota A, Rouvier J, Gonzalez C. Effects of ticlopidine or ticlopidine plus aspirin on platelet aggregation and ATP release in normal volunteers: why aspirin improves ticlopidine antiplatelet activity. *Clin Appl Thromb Hemost*. 1999;5(4):243−246.

20. Helgason CM, Bolin KM, Hoff JA, et al. Development of aspirin resistance in persons with previous ischemic stroke. *Stroke*. 1994;25(12):2331−2336.

21. Serruys PW, Strauss BH, Beatt KJ, et al. Angiographic follow-up after placement of a self-expanding coronary-artery stent. *N Engl J Med*. 1991;324(1):13−17.

22. Colombo A, Hall P, Nakamura S, et al. Intracoronary stenting without anticoagulation accomplished with intravascular ultrasound guidance. *Circulation*. 1995;91(6):1676−1688.

23. Hall P, Nakamura S, Maiello L, et al. A randomized comparison of combined ticlopidine and aspirin therapy versus aspirin therapy alone after successful intravascular ultrasound-guided stent implantation. *Circulation*. 1996;93(2):215−222.

24. Leon MB, Baim DS, Popma JJ, et al. A clinical trial comparing three antithrombotic-drug regimens after coronary-artery stenting. Stent Anticoagulation Restenosis Study Investigators. *N Engl J Med*. 1998;339(23):1665−1671.

25. Schomig A, Neumann FJ, Kastrati A, et al. A randomized comparison of antiplatelet and anticoagulant therapy after the placement of coronary-artery stents. *N Engl J Med*. 1996;334(17):1084−1089.

26. Mehta SJL, Clopidogrel in Unstable Angina to Prevent Recurrent Events Trial (CURE) Investigators. Effects of pretreatment with clopidogrel and aspirin followed by long-term therapy in patients undergoing percutaneous coronary intervention: the PCI-CURE study. *Lancet*. 2001;358:527−533.

27. Neumann F-J, Sousa-Uva M, Ahlsson A, et al. 2018 ESC/EACTS guidelines on myocardial revascularization. *Kardiol Pol*. 2018;40(2):87−165.

28. Yusuf S, Zhao F, Mehta S, Chrolavicius S, Tognoni GJNEJM, Fox KK, Clopidogrel in Unstable Angina to Prevent Recurrent Events Trial Investigators. Effects of clopidogrel in addition to aspirin in patients with acute coronary syndromes without ST-segment elevation. *N Engl J Med*. 2001;345:494−502.

29. Cook S, Windecker S. Early stent thrombosis: past, present, and future. *Circulation*. 2009;119(5):657−659.

30. Kandzari DE, Leon MB, Meredith I, Fajadet J, Wijns W, Mauri L. Final 5-year outcomes from the Endeavor zotarolimus-eluting stent clinical trial program: comparison of safety and efficacy with first-generation drug-eluting and bare-metal stents. *JACC Cardiovasc Interv*. 2013;6(5):504−512.

31. Kinnaird TD, Stabile E, Mintz GS, et al. Incidence, predictors, and prognostic implications of bleeding and blood transfusion following percutaneous coronary interventions. *Am J Cardiol*. 2003;92(8):930−935.

32. Valgimigli M, Campo G, Monti M, et al. Short- versus long-term duration of dual-antiplatelet therapy after coronary stenting: a randomized multicenter trial. *Circulation*. 2012;125(16):2015−2026.

33. Gwon HC, Hahn JY, Park KW, et al. Six-month versus 12-month dual antiplatelet therapy after implantation of drug-eluting stents: the Efficacy of Xience/Promus versus Cypher to Reduce Late Loss After Stenting (EXCELLENT) randomized, multicenter study. *Circulation*. 2012;125(3):505−513.

34. Colombo A, Chieffo A, Frasheri A, et al. Second-generation drug-eluting stent implantation followed by 6-versus 12-month dual antiplatelet therapy the SECURITY randomized clinical trial. *J Am Coll Cardiol*. 2014;64(20):2086−2097.

35. Didier R, Morice MC, Barragan P, et al. 6-Versus 24-month dual antiplatelet therapy after implantation of drug-eluting stents in patients nonresistant to aspirin final results of the ITALIC trial (is there a life for DES after discontinuation of clopidogrel). *JACC Cardiovasc Interv*. 2017;10(12):1202−1210.

36. Schulz-Schupke S, Byrne RA, ten Berg JM, et al. ISAR-SAFE: a randomized, double-blind, placebo-controlled trial of 6 vs. 12 months of clopidogrel therapy after drug-eluting stenting. *Eur Heart J*. 2015;36(20):1252−1263.

37. Han Y, Xu B, Xu K, et al. Six versus 12 Months of dual antiplatelet therapy after implantation of biodegradable polymer sirolimus-eluting stent: randomized substudy of the I-LOVE-IT 2 trial. *Circ Cardiovasc Interv*. 2016;9(2):e003145.

38. Feres F, Costa RA, Bhatt DL, et al. Optimized duration of clopidogrel therapy following treatment with the Endeavor zotarolimus-eluting stent in real-world clinical practice (OPTIMIZE) trial: rationale and design of a large-scale, randomized, multicenter study. *Am Heart J*. 2012;164(6):810−816. e813.

39. Kim BK, Hong MK, Shin DH, et al. A new strategy for discontinuation of dual antiplatelet therapy: the RESET Trial (REal Safety and Efficacy of 3-month dual antiplatelet Therapy following Endeavor zotarolimus-eluting stent implantation). *J Am Coll Cardiol*. 2012;60(15):1340−1348.

40. Collet JP, Silvain J, Barthelemy O, et al. Dual-antiplatelet treatment beyond 1 year after drug-eluting stent implantation (ARCTIC-Interruption): a randomised trial. *Lancet*. 2014;384(9954):1577−1585.

41. Mauri L, Kereiakes DJ, Yeh RW, et al. Twelve or 30 months of dual antiplatelet therapy after drug-eluting stents. *N Engl J Med*. 2014;371(23):2155−2166.

42. Lee CW, Ahn JM, Park DW, et al. Optimal duration of dual antiplatelet therapy after drug-eluting stent implantation: a randomized, controlled trial. *Circulation*. 2014;129(3):304−312.

43. Helft G, Steg PG, Le Feuvre C, et al. Stopping or continuing clopidogrel 12 months after drug-eluting stent placement: the OPTIDUAL randomized trial. *Eur Heart J*. 2016;37(4):365−374.

44. Navarese EP, Andreotti F, Schulze V, et al. Optimal duration of dual antiplatelet therapy after percutaneous coronary intervention with drug eluting stents: meta-analysis of randomised controlled trials. *BMJ*. 2015;350. h1618-h1618.

45. De Berardis G, Lucisano G, D'Ettorre A, et al. Association of aspirin use with major bleeding in patients with and without diabetes. *JAMA*. 2012;307(21):2286—2294.

46. McNeil JJ, Wolfe R, Woods RL, et al. Effect of aspirin on cardiovascular events and bleeding in the healthy elderly. *N Engl J Med*. 2018;379(16):1509—1518.

47. Antithrombotic Trialists C, Baigent C, Blackwell L, et al. Aspirin in the primary and secondary prevention of vascular disease: collaborative meta-analysis of individual participant data from randomised trials. *Lancet*. 2009;373(9678):1849—1860.

48. Weisman SM, Graham DY. Evaluation of the benefits and risks of low-dose aspirin in the secondary prevention of cardiovascular and cerebrovascular events. *Arch Intern Med*. 2002;162(19):2197—2202.

49. Desborough MJ, Keeling DM. The aspirin story—from willow to wonder drug. *Br J Haematol*. 2017;177(5):674—683.

50. Sneader W. The discovery of aspirin: a reappraisal. *BMJ*. 2000;321(7276):1591—1594.

51. Weiss HJ. The discovery of the antiplatelet effect of aspirin: a personal reminiscence. *J Thromb Haemost*. 2003;1(9):1869—1875.

52. Patrono C, Garcia Rodriguez LA, Landolfi R, Baigent C. Low-dose aspirin for the prevention of atherothrombosis. *N Engl J Med*. 2005;353(22):2373—2383.

53. Graham MM, Sessler DI, Parlow JL, et al. Aspirin in patients with previous percutaneous coronary intervention undergoing noncardiac surgery. *Ann Intern Med*. 2018; 168(4):237—244.

54. Sundstrom J, Hedberg J, Thuresson M, Aarskog P, Johannesen KM, Oldgren J. Low-dose aspirin discontinuation and risk of cardiovascular events a Swedish Nationwide, population-based cohort study. *Circulation*. 2017;136(13):1183—+.

55. Force USPST. Routine aspirin or nonsteroidal anti-inflammatory drugs for the primary prevention of colorectal cancer: U.S. Preventive Services Task Force recommendation statement. *Ann Intern Med*. 2007;146(5):361—364.

56. Franklin BA. Aspirin for the primary prevention of cardiovascular events: considerations regarding the risk/benefit. *Phys Sportsmed*. 2010;38(1):158—161.

57. Roffi M, Patrono C, Collet JP, et al. 2015 ESC guidelines for the management of acute coronary syndromes in patients presenting without persistent ST-segment elevation Task Force for the management of acute coronary syndromes in patients presenting without persistent ST-segment elevation of the European Society of Cardiology (ESC). *Eur Heart J*. 2016;37(3):267.

58. Ibanez B, James S, Agewall S, et al, ESC Scientific Document Group. 2017 ESC Guidelines for the management of acute myocardial infarction in patients presenting with ST-segment elevation: The Task Force for the management of acute myocardial infarction in patients presenting with ST-segment elevation of the European Society of Cardiology (ESC). *Eur Heart J*. 2018;39(2):119—177.

59. Piepoli M, Hoes AJ, Agewall S, et al. European guidelines on cardiovascular disease prevention in clinical practice: the sixth joint task force of the European Society of Cardiology and other societies on cardiovascular disease prevention in clinical practice (constituted by representatives of 10 societies and by invited experts) developed with the special contribution of the European Association for Cardiovascular Prevention & Rehabilitation (EACPR). *Eur Heart J*. 2016;37(29):2315—2381.

60. Gerhard-Herman MD, Gornik HL, Barrett C, et al. 2016 AHA/ACC guideline on the management of patients with lower extremity peripheral artery disease: a report of the American College of Cardiology/American Heart Association Task Force on Clinical Practice Guidelines. *J Am Coll Cardiol*. 2017;69(11):e71–e126.

61. Gargiulo G, Capodanno D, Longo G, Capranzano P, Tamburino CJE. Updates on NSAIDs in patients with and without coronary artery disease: pitfalls, interactions and cardiovascular outcomes. *Expert Rev Cardiovasc Ther*. 2014;12(10):1185–1203.

62. Whitlock EP, Burda BU, Williams SB, Guirguis-Blake JM, Evans CV. Bleeding risks with aspirin use for primary prevention in adults: a systematic review for the US Preventive Services Task Force. *Ann Intern Med*. 2016;164(12):826–U892.

63. Chen WC, Lin KH, Huang YT, et al. The risk of lower gastrointestinal bleeding in low-dose aspirin users. *Aliment Pharmacol Ther*. 2017;45(12):1542–1550.

64. Huang WY, Saver JL, Wu YL, Lin CJ, Lee M, Ovbiagele B. Frequency of intracranial hemorrhage with low-dose aspirin in individuals without Symptomatic cardiovascular disease: a systematic review and meta-analysis. *JAMA Neurol*. 2019;76(8):906–914.

65. Phan K, Moore JM, Griessenauer CJ, Ogilvy CS, Thomas AJ. Aspirin and risk of subarachnoid hemorrhage: systematic review and meta-analysis. *Stroke*. 2017;48(5):1210–1217.

66. Gaziano JM, Brotons C, Coppolecchia R, et al. Use of aspirin to reduce risk of initial vascular events in patients at moderate risk of cardiovascular disease (ARRIVE): a randomised, double-blind, placebo-controlled trial. *Lancet*. 2018;392(10152):1036–1046.

67. Medicine ASCGJNEJo. Effects of aspirin for primary prevention in persons with diabetes mellitus. *N Engl J Med*. 2018;379(16):1529–1539.

68. McNeil JJ, Nelson MR, Woods RL, et al. Effect of aspirin on all-cause mortality in the healthy elderly. *N Engl J Med*. 2018;379(16):1519–1528.

69. Mahmoud AN, Gad MM, Elgendy AY, Elgendy IY, Bavry AA. Efficacy and safety of aspirin for primary prevention of cardiovascular events: a meta-analysis and trial sequential analysis of randomized controlled trials. *Eur Heart J*. 2019;40(7):607–617.

70. Baigent C, Blackwell L, Collins R, et al. *Aspirin in the primary and secondary prevention of vascular disease: collaborative meta-analysis of individual participant data from randomised trials*. Elsevier; 2009.

71. Baigent C, Blackwell L, Collins R, et al. Antithrombotic Trialists'(ATT) Collaboration. Aspirin in the primary and secondary prevention of vascular disease: collaborative meta-analysis of individual participant data from randomised trials. *Lancet*. 2009;373(9678):1849–1860.

72. Vane JR, Botting RM. The mechanism of action of aspirin. *Thromb Res*. 2003;110(5–6):255–258.

73. Wallentin L. P2Y$_{12}$ inhibitors: differences in properties and mechanisms of action and potential consequences for clinical use. *Eur Heart J*. 2009;30(16):1964–1977.

74. Cadroy Y, Bossavy JP, Thalamas C, Sagnard L, Sakariassen K, Boneu B. Early potent antithrombotic effect with combined aspirin and a loading dose of clopidogrel on experimental arterial thrombogenesis in humans. *Circulation*. 2000;101(24):2823–2828.

75. Jakubowski JA, Winters KJ, Naganuma H, Wallentin L. Prasugrel: a novel thienopyridine antiplatelet agent. A review of preclinical and clinical studies and the mechanistic basis for its distinct antiplatelet profile. *Cardiovasc Drug Rev*. 2007;25(4):357–374.

76. Angiolillo DJ. Variability in responsiveness to oral antiplatelet therapy. *Am J Cardiol*. 2009;103(3 Suppl):27A–34A.

77. Jennings LK. Role of platelets in atherothrombosis. *Am J Cardiol.* 2009;103(3 Suppl): 4A—10A.
78. Li Z, Zhang G, Le Breton GC, Gao X, Malik AB, Du X. Two waves of platelet secretion induced by thromboxane A2 receptor and a critical role for phosphoinositide 3-kinases. *J Biol Chem.* 2003;278(33):30725—30731.
79. Paul BZ, Jin J, Kunapuli SP. Molecular mechanism of thromboxane A2-induced platelet aggregation ESSENTIAL ROLE FOR P2T AC and α2ARECEPTORS. *J Biol Chem.* 1999;274(41):29108—29114.
80. Armstrong PC, Dhanji AR, Tucker AT, Mitchell JA, Warner TD. Reduction of platelet thromboxane A2 production ex vivo and in vivo by clopidogrel therapy. *J Thromb Haemost.* 2010;8(3):613—615.
81. Rollini F, Franchi F, Cho JR, et al. Cigarette smoking and antiplatelet effects of aspirin monotherapy versus clopidogrel monotherapy in patients with atherosclerotic disease: results of a prospective pharmacodynamic study. *J Cardiovasc Transl Res.* 2014;7(1): 53—63.
82. Angiolillo DJ, Capranzano P, Desai B, et al. Impact of $P2Y_{12}$ inhibitory effects induced by clopidogrel on platelet procoagulant activity in type 2 diabetes mellitus patients. *Thromb. Res.* 2009;124(3):318—322.
83. Bhavaraju K, Georgakis A, Jin J, et al. Antagonism of $P2Y_{12}$ reduces physiological thromboxane levels. *Platelets.* 2010;21(8):604—609.
84. Björkman J-A, Zachrisson H, Forsberg G-B, et al. High-dose aspirin in dogs increases vascular resistance with limited additional anti-platelet effect when combined with potent $P2Y_{12}$ inhibition. *Thromb. Res.* 2013;131(4):313—319.
85. Kirkby N, Leadbeater P, Chan M, et al. Antiplatelet effects of aspirin vary with level of $P2Y_{12}$ receptor blockade supplied by either ticagrelor or prasugrel. *J Thromb Haemost.* 2011;9(10):2103—2105.
86. Armstrong PC, Leadbeater PD, Chan MV, et al. In the presence of strong $P2Y_{12}$ receptor blockade, aspirin provides little additional inhibition of platelet aggregation. *J Thromb Haemost.* 2011;9(3):552—561.
87. Storey RF, Sanderson HM, White AE, May JA, Cameron KE, Heptinstall S. The central role of the P_{2T} receptor in amplification of human platelet activation, aggregation, secretion and procoagulant activity. *Br J Haematol.* 2000;110(4):925—934.
88. Hass WK, Easton JD, Adams Jr HP, et al. A randomized trial comparing ticlopidine hydrochloride with aspirin for the prevention of stroke in high-risk patients. Ticlopidine Aspirin Stroke Study Group. *N Engl J Med.* 1989;321(8):501—507.
89. Lancet CSCJT. A randomised, blinded, trial of clopidogrel versus aspirin in patients at risk of ischaemic events (CAPRIE). *Lancet.* 1996;348(9038):1329—1339.
90. Diener H-C, Bogousslavsky J, Brass LM, et al. Aspirin and clopidogrel compared with clopidogrel alone after recent ischaemic stroke or transient ischaemic attack in high-risk patients (MATCH): randomised, double-blind, placebo-controlled trial. *Lancet.* 2004; 364(9431):331—337.
91. Stone GW, Witzenbichler B, Weisz G, et al. Platelet reactivity and clinical outcomes after coronary artery implantation of drug-eluting stents (ADAPT-DES): a prospective multicentre registry study. *Lancet.* 2013;382(9892):614—623.
92. Stuckey TD, Kirtane AJ, Brodie BR, et al. Impact of aspirin and clopidogrel hyporesponsiveness in patients treated with drug-eluting stents: 2-year results of a prospective, multicenter registry study. *JACC Cardiovasc Interv.* 2017;10(16):1607—1617.

93. Yano M, Natsuaki M, Morimoto T, et al. Antiplatelet therapy discontinuation and stent thrombosis after sirolimus-eluting stent implantation: five-year outcome of the j-Cypher Registry. *Int J Cardiol.* 2015;199:296—301.

94. Watanabe H, Morimoto T, Natsuaki M, et al. Antiplatelet therapy discontinuation and the risk of serious cardiovascular events after coronary stenting: observations from the CREDO-Kyoto Registry Cohort-2. *PLoS One.* 2015;10(4):e0124314.

95. Lopes RD, Heizer G, Aronson R, et al. Antithrombotic therapy after acute coronary syndrome or PCI in atrial fibrillation. *N Engl J Med.* 2019;380(16):1509—1524.

96. Lopes RD, Hong H, Harskamp RE, et al. Safety and efficacy of antithrombotic strategies in patients with atrial fibrillation undergoing percutaneous coronary intervention: a network meta-analysis of randomized controlled trials. *JAMA Cardiol.* 2019;4(8):747—755.

97. Committee CS. A randomised, blinded, trial of clopidogrel versus aspirin in patients at risk of ischaemic events (CAPRIE). CAPRIE Steering Committee. *Lancet.* 1996; 348(9038):1329—1339.

98. Hahn JY, Song YB, Oh JH, et al. Effect of P2Y12 inhibitor monotherapy vs dual antiplatelet therapy on cardiovascular events in patients undergoing percutaneous coronary intervention: the SMART-CHOICE randomized clinical trial. *J Am Med Assoc.* 2019;321(24):2428—2437.

99. Ziada KM, Moliterno DJ. Dual antiplatelet therapy: is it time to cut the cord with aspirin? *J Am Med Assoc.* 2019;321(24):2409—2411.

100. Watanabe H, Domei T, Morimoto T, et al. Effect of 1-month dual antiplatelet therapy followed by clopidogrel vs 12-month dual antiplatelet therapy on cardiovascular and bleeding events in patients receiving PCI: the STOPDAPT-2 randomized clinical trial. *J Am Med Assoc.* 2019;321(24):2414—2427.

101. Levine GN, Jeong YH, Goto S, et al. Expert consensus document: world Heart Federation expert consensus statement on antiplatelet therapy in East Asian patients with ACS or undergoing PCI. *Nat Rev Cardiol.* 2014;11(10):597—606.

102. Jeong Y-H. "East asian paradox": challenge for the current antiplatelet strategy of "one-guideline-fits-all races" in acute coronary syndrome. *Curr Cardiol Rep.* 2014;16(5): 485.

103. Xie HG, Kim RB, Wood AJ, Stein CM. Molecular basis of ethnic differences in drug disposition and response. *Annu Rev Pharmacol Toxicol.* 2001;41(1):815—850.

104. Vranckx P, Valgimigli M, Windecker S, et al. Long-term ticagrelor monotherapy versus standard dual antiplatelet therapy followed by aspirin monotherapy in patients undergoing biolimus-eluting stent implantation: rationale and design of the GLOBAL LEADERS trial. *EuroIntervention.* 2016;12(10):1239—1245.

105. Vranckx P, Valgimigli M, Juni P, et al. Ticagrelor plus aspirin for 1 month, followed by ticagrelor monotherapy for 23 months vs aspirin plus clopidogrel or ticagrelor for 12 months, followed by aspirin monotherapy for 12 months after implantation of a drug-eluting stent: a multicentre, open-label, randomised superiority trial. *Lancet.* 2018;392(10151):940—949.

106. Franzone A, McFadden E, Leonardi S, et al. Ticagrelor alone versus dual antiplatelet therapy from 1 month after drug-eluting coronary stenting. *J Am Coll Cardiol.* 2019; 74(18):2223—2234.

107. Mehran R, Baber U, Sharma SK, et al. Ticagrelor with or without aspirin in high-risk patients after PCI. *N Engl J Med.* 2019;381(21):2032—2042.

108. Baber U, Zafar MU, Dangas G, et al. Ticagrelor with or without aspirin after PCI: The TWILIGHT platelet substudy. *J Am Coll Cardiol.* 2020;75(6):578—586.

109. Kim BK, Hong SJ, Shin DH, et al. Effect of ticagrelor monotherapy vs ticagrelor with aspirin on major bleeding and cardiovascular events in patients with acute coronary syndrome: The TICO randomized clinical trial. *JAMA*. 2020;323:2407–2416.
110. Kogame N, Modolo R, Tomaniak M, et al. Prasugrel monotherapy after PCI with the SYNERGY stent in patients with chronic stable angina or stabilised acute coronary syndromes: rationale and design of the ASET pilot study. *EuroIntervention*. 2019;15(6): e547–e550.
111. Urban P, Mehran R, Colleran R, et al. Defining high bleeding risk in patients undergoing percutaneous coronary intervention: a consensus document from the Academic Research Consortium for High Bleeding Risk. *Eur Heart J*. 2019;40(31):2632–2653.

The conundrum of simultaneous antiplatelet and anticoagulant therapy: how to solve it?

11

Freek W.A. Verheugt, MD, PhD [1], **Robert F. Storey, MD** [2]

[1]*Department of Cardiology, Heart Center, Onze Lieve Vrouwe Gasthuis (OLVG), Amsterdam, Netherlands;* [2]*Department of Infection, Immunity and Cardiovascular Disease, University of Sheffield, Sheffield, United Kingdom*

Antiplatelet therapy is the cornerstone of secondary prevention of coronary artery disease (CAD), and dual antiplatelet therapy (DAPT) has become the standard of care of patients undergoing percutaneous coronary intervention (PCI) with coronary stenting and of patients with acute coronary syndromes (ACSs) with or without stent implantation. In patients with non-ST-segment elevation ACS, 1 year of treatment is recommended.[1] In elective patients, 6 months of DAPT is preferred by most cardiologists, but prolonged treatment can be given with more complex anatomy or the use of multiple stents. Thus many stented patients are on DAPT, mainly aspirin and clopidogrel. The only serious side effect of DAPT is increased bleeding in comparison to aspirin alone.[2,3]

Up to 10% of patients with CAD have concomitant atrial fibrillation (AF), of whom 3%–5% will undergo coronary stenting.[4] Oral anticoagulation (OAC) with a vitamin K antagonist (VKA) or non-VKA oral anticoagulant (NOAC) is mandatory for the prevention of ischemic stroke in these patients. Needless to say, the combination of antiplatelets and OAC increases the risk of bleeding further. When DAPT is combined with warfarin, bleeding increased twofold to threefold in clinical trials[5,6] as well as in a large Danish registry.[7]

Options to reduce the bleeding risk of combined antiplatelet and anticoagulant therapy

Radial access

In interventional cardiology, a routine radial approach in a wide range of CAD presentations has been shown to be superior to femoral access with regard to safety, not only in terms of reduced vascular complications but also potentially lower mortality.[8] Although this has not been prospectively tested, it is very likely that this holds true for anticoagulated patients with atrial fibrillation.

Dual Antiplatelet Therapy for Coronary and Peripheral Arterial Disease
https://doi.org/10.1016/B978-0-12-820536-5.00012-4

Various use of P2Y$_{12}$ inhibitors

As mentioned earlier, DAPT is mandatory after coronary stenting. When shortening the duration of DAPT from 6 months to 6 weeks, a significant reduction of major bleeding by 32% was achieved in 614 anticoagulated patients in the randomized ISAR-TRIPLE trial.[9] Depending on the indication for intervention, clopidogrel for either elective PCI or PCI in ACS, the stronger P2Y$_{12}$ inhibitors ticagrelor or prasugrel are used, all in combination with aspirin. Both prasugrel and ticagrelor have been shown to increase bleeding relative to clopidogrel.[10,11] In a small observational study, prasugrel seemed to quadruple bleeding in anticoagulated patients when compared to clopidogrel, but without reduction of ischemic outcomes.[12] In a much larger substudy of the RE-DUAL PCI trial (see later discussion), the use of ticagrelor in anticoagulated patients with AF neither significantly increased bleeding in comparison with clopidogrel nor reduced ischemic events.[13] Yet, randomized head-to-head studies comparing the novel P2Y$_{12}$ inhibitors in anticoagulated patients are lacking. In view of safety concerns, their routine use should be avoided in anticoagulated patients with AF undergoing PCI for ACS when a combination of DAPT and OAC is used.[1] Only in patients with a very high ischemic risk (e.g., refractory coronary thrombosis) and a low bleeding risk may the stronger P2Y$_{12}$ inhibitors be applied in a triple antithrombotic combination.

Dropping aspirin

The first study to prospectively evaluate the removal of aspirin from triple therapy (OAC plus DAPT) was the WOEST trial.[14] This randomized trial, in 576 patients anticoagulated with VKA undergoing PCI, showed that single antiplatelet therapy with clopidogrel without aspirin in anticoagulated patients was safer than triple therapy by reducing bleeding by 64%, without increasing ischemic events.

Non—vitamin K antagonist oral anticoagulants

For stroke prevention, NOACs are noninferior to warfarin and also safer in use, except for gastrointestinal (GI) bleeding.[15] In general, current guidelines recommend the use of NOACs in preference to VKA in NOAC-eligible patients with AF undergoing PCI[16–18] on the basis of the four randomized controlled studies that have compared NOACs with VKA in this population.[19–22] In most of the studies, the patients randomized to an NOAC were not given aspirin (as was done with VKA in WOEST[15]), whereas aspirin was administered to the patients randomized to VKA, albeit for a period from only 1 to up to 12 months at the discretion of the randomizing physician. Bleeding was the primary endpoint in these studies. The bottom-line results are that for stroke prevention, NOACs are at least not inferior to VKA after PCI, in terms of prevention of ischemic events, and are much safer. For the latter finding, one should realize that dual therapy (OAC plus a P2Y$_{12}$ inhibitor) is expected to be safer than triple therapy (OAC, P2Y$_{12}$ inhibitor, and aspirin). So that seems an easy winner. Furthermore, in PIONEER AF-PCI, two doses of the

NOAC (rivaroxaban) were used that were lower than the approved full dose for AF: 15 mg qd and 2.5 mg bid.[19] Only the AUGUSTUS trial had a better setup. A factorial design was used: apixaban versus VKA (open label) and aspirin versus placebo (double-blind).[21] Bleeding was lowest on apixaban without aspirin and highest with VKA with aspirin (Fig. 11.1). Ischemic endpoints were quite similar in the four arms in that trial, which was the largest of its kind. There was a nonsignificant increase in stent thrombosis in the arms without aspirin, but this finding did not lead to significantly more myocardial infarctions.[23,24]

Use of proton pump inhibitors

With increasing numbers of elderly people, the incidence of AF is increasing and also bleeding is becoming more common. In the elderly, GI bleeding is the most common form of major and sometimes fatal bleeding. Especially, aspirin is associated with an increased risk of GI bleeding as found in secondary prevention of vascular disease.[25] To prevent gastric ulceration by aspirin, the use of proton pump inhibitors (PPIs) has been shown to be effective.[26] Also, GI events can be reduced by PPIs in patients on DAPT after PCI.[27] In that placebo-controlled study, no interaction was found between clopidogrel and omeprazole. For patients on oral anticoagulants, PPIs might reduce bleeding; however, randomized trials in this field are lacking. PPIs do not seem to interact with NOACs but may interfere with INR (International Normalized Ratio) in patients on warfarin, requiring care in warfarin dosing and INR management.[28]

In patients without an indication for OAC, the use of PPIs has been evaluated in the substudy of COMPASS on 17,598 patients. COMPASS was a very large clinical trial evaluating low-dose rivaroxaban and very-low-dose rivaroxaban on top of aspirin versus aspirin alone in stable vascular patients. In the PPI substudy,

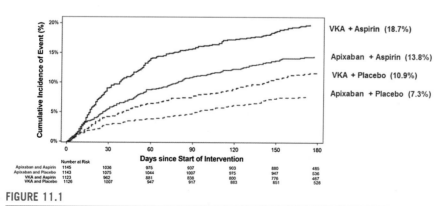

FIGURE 11.1

Major bleeding and clinically relevant nonmajor bleeding in the AUGUSTUS trial.[21] *VKA, vitamin K antagonist.*

Reproduced with permission from Renato Lopes, Duke University, Durham, USA.

pantoprazole reduced gastroduodenal events by 42% in the patients on aspirin, but not in those on rivaroxaban.[29] Yet, the doses of rivaroxaban in that study were much lower than those used for stroke prevention in AF.

Avoidance of nonsteroidal anti-inflammatory drugs

Anti-inflammatory agents such as nonsteroidal anti-inflammatory drugs (NSAIDs) are commonly used for relief of many forms of pain, such as postoperative pain or chronic pain in rheumatoid disorders. Many of these agents interfere with renal function, and also cause bleeding. In a large Danish registry of over 150,000 patients with AF on OAC, 3 months' use of nonaspirin NSAIDs was associated with a significant increase in serious bleeding: 2.5 pts per 1000 pts treated.[30] Selective COX-2 inhibitors (coxibs) are no alternative for these patients, given their increased risk of ischemic events, but paracetamol as a painkiller may be useful, as recommended by the expert opinion paper of the European Society of Cardiology.[31]

The options to reduce the bleeding risk of combined antiplatelet and anticoagulant therapy and their potential implementation are summarized in Table 11.1.

Future perspectives

Antiplatelet agents will always be necessary in patients with vascular disease. For patients on simultaneous antiplatelet and OAC, monotherapy with a $P2Y_{12}$ inhibitor without aspirin has become a reality. A new $P2Y_{12}$ inhibitor on the horizon is selatogrel, a subcutaneous agent that has rapid onset of action and appears well tolerated.[32] Its clinical applications are still to be explored and its safety in combination with OAC has not yet been assessed. Another new antiplatelet agent is the glycoprotein IIb/IIIa receptor antagonist RUC-4, which has been developed for intramuscular or subcutaneous use.[33] It is currently evaluated in patients with acute myocardial infarction, and its role in anticoagulated patients remains to be established.

Table 11.1 Options to reduce bleeding in simultaneous antiplatelet and anticoagulant therapy.

Factor increasing bleeding	Alternative
Femoral access	Radial access
Use of aspirin	Drop aspirin
DAPT with ticagrelor/prasugrel	Clopidogrel
Vitamin K antagonist	NOAC
NSAIDs	Paracetamol

DAPT, *dual antiplatelet therapy*; NOAC, *non–vitamin K antagonist oral anticoagulant*; NSAIDs, *nonsteroidal anti-inflammatory drugs.*

Finally, new oral anticoagulants blocking the activity of activated factor XI, a procoagulant substance in the intrinsic pathway of the coagulation cascade, have been developed. These agents will be evaluated for use in stroke prevention in patients with AF as well as in the prevention and/or treatment of venous thromboembolism.[34] There is no experience of this treatment modality in patients requiring antiplatelet therapy.

Conclusion

To reduce bleeding in patients on simultaneous antiplatelet and anticoagulant therapy, combining OAC and antiplatelet monotherapy with a $P2Y_{12}$ inhibitor without aspirin has become a reality. Furthermore, the wide application of NOACs in patients with AF has made PCI safer. Each physician taking care of these patients should, whenever feasible, use radial access; avoid aspirin as an antiplatelet drug; avoid NSAIDs; and administer PPIs to patients at high risk of GI bleeding.

Conflicts of interest

FWAV has received educational and research grants from Bayer Healthcare and Boehringer-Ingelheim, as well as honoraria for consultancies/presentations from Daiichi-Sankyo, BMS/Pfizer, Boehringer-Ingelheim, and Bayer Healthcare.

RFS reports research grants, consultancy fees, and honoraria from AstraZeneca; consultancy fees and honoraria from Bayer and Bristol Myers Squibb/Pfizer; research grants and consultancy fees from GlyCardial Diagnostics and Thromboserin; consultancy fees from Amgen, Haemonetics, and Portola; and honoraria from Medscape.

References

1. Verheugt FWA, ten Berg JM, Storey RF, Cuisset T, Granger CB. Antithrombotics: from aspirin to DOACs in coronary artery disease and atrial fibrillation. *J Am Coll Cardiol.* 2019;74:699−711.
2. CURE Investigators. Effect of clopidogrel in addition to aspirin in patients with acute coronary syndromes without ST-segment elevation. *N Engl J Med.* 2001;345:494−502.
3. COMMIT (Clopidogrel and Metoprolol in Myocardial Infarction Trial) Collaborative Group. Addition of clopidogrel to aspirin in 45852 patients with acute myocardial infarction: randomised placebo-controlled trial. *Lancet.* 2005;366:1607−1621.
4. Hansen ML, Sørensen R, Clausen MT, et al. Risk of bleeding with single, dual, or triple therapy with warfarin, aspirin, and clopidogrel in patients with atrial fibrillation. *Arch Intern Med.* 2010;170:1433−1441.
5. Dans AL, Connolly SJ, Wallentin L, et al. Concomitant use of antiplatelet therapy with dabigatran or warfarin in the Randomized Evaluation of Long-Term Anticoagulation Therapy (RE-LY) trial. *Circulation.* 2013;127:634−640.

6. Alexander JH, Lopes RD, Thomas L, et al. Apixaban vs. warfarin with concomitant aspirin in patients with atrial fibrillation: insights from the ARISTOTLE trial. *Eur Heart J*. 2014;35:224–232.
7. Lamberts M, Olesen JB, Ruwald MH, et al. Bleeding after initiation of multiple antithrombotic drugs, including triple therapy, in atrial fibrillation patients following myocardial infarction and coronary intervention: a nationwide cohort study. *Circulation*. 2012; 126:1185–1293.
8. Ferrante G, Rao SV, Jüni P, et al. Radial versus femoral access for coronary interventions across the entire spectrum of patients with coronary artery disease: a meta-analysis of randomized trials. *JACC Cardiovasc Interv*. 2016;9:1419–1934.
9. Fiedler KA, Maeng M, Mehilli J, et al. Duration of triple therapy in patients requiring oral anticoagulation after drug-eluting stent implantation: the ISAR-TRIPLE trial. *J Am Coll Cardiol*. 2015;65:1619–1629.
10. Wiviott SD, Braunwald E, McCabe CH, et al. Prasugrel versus clopidogrel in patients with acute coronary syndromes. *N Engl J Med*. 2007;357:2001–2015.
11. Wallentin L, Becker RC, Budaj A, et al. Ticagrelor versus clopidogrel in patients with acute coronary syndromes. *N Engl J Med*. 2009;361:1045–1057.
12. Sarafoff N, Martischnig A, Wealer J, et al. Triple therapy with aspirin, prasugrel, and vitamin K antagonists in patients with drug-eluting stent implantation and an indication for oral anticoagulation. *J Am Coll Cardiol*. 2013;61:2060–2066.
13. Oldgren J, Steg PG, Hohnloser SH, et al. Dabigatran dual therapy with ticagrelor or clopidogrel after percutaneous coronary intervention in atrial fibrillation patients with or without acute coronary syndrome: a subgroup analysis from the RE-DUAL PCI trial. *Eur Heart J*. 2019;40:1553–1562.
14. Dewilde WJM, Oirbans T, Verheugt FWA, et al. Use of clopidogrel with or without aspirin in patients taking oral anticoagulant therapy and undergoing percutaneous coronary intervention: an open-label, randomised controlled trial. *Lancet*. 2013;381: 1107–1115.
15. Ruff CT, Giugliano RP, Braunwald E, et al. Comparison of the efficacy and safety of new oral anticoagulants with warfarin in patients with atrial fibrillation: a meta-analysis of randomised trials. *Lancet*. 2014;383:955–962.
16. Kirchhof P, Benussi S, Kotecha D, et al. 2016 ESC guidelines for the management of atrial fibrillation developed in collaboration with the EACTS. *Eur Heart J*. 2016;37: 2893–2962.
17. Knuuti J, Wijns W, Saraste A, et al. 2019 ESC Guidelines for the diagnosis and management of chronic coronary syndromes. *Eur Heart J*. 2019;39:213–260. Online 2018.
18. Angiolillo DJ, Goodman SH, Bhatt DL, et al. Antithrombotic therapy in patients with atrial fibrillation treated with oral anticoagulation undergoing percutaneous coronary intervention: a North American perspective 2018 update. *Circulation*. 2018;138: 527–536.
19. Gibson CM, Mehran R, Bode C, et al. Prevention of bleeding in patients with atrial fibrillation undergoing PCI. *N Engl J Med*. 2016;375:2423–2434.
20. Cannon CP, Bhatt DL, Oldgren J, et al. Dual antithrombotic therapy with dabigatran after PCI in atrial fibrillation. *N Engl J Med*. 2017;377:1513–1524.
21. Lopes RD, Heizer G, Aronson R, et al. Antithrombotic therapy after acute coronary syndrome or PCI in atrial fibrillation. *N Engl J Med*. 2019;380:1509–1524.
22. Vranckx P, Valgimigli M, Eckardt L, et al. Edoxaban-based versus vitamin K antagonist-based antithrombotic regimen after successful coronary stenting in patients with atrial

fibrillation (ENTRUST-AF PCI): a randomised, open-label, phase 3b trial. *Lancet.* 2019; 394:1335−1343.

23. Lopes RD, Leonardi S, Wojdyla DM, et al. Stent thrombosis in patients with atrial fibrillation undergoing coronary stenting in the AUGUSTUS Trial. *Circulation.* November 11, 2019. https://doi.org/10.1161/CIRCULATIONAHA.119.044584.

24. Gargiulo G, Goette A, Tijssen J, et al. Safety and efficacy outcomes of double vs. triple antithrombotic therapy in patients with atrial fibrillation following percutaneous coronary intervention: a systematic review and meta-analysis of non-vitamin K antagonist oral anticoagulant-based randomized clinical trials. *Eur Heart J.* December 7, 2019; 40(46):3757−3767. https://doi.org/10.1093/eurheartj/ehz732. PMID: 31651946.

25. Li L, Geraghty OC, Mehta Z, Rothwell PM. Age-specific risks, severity, time course, and outcome of bleeding on long-term antiplatelet treatment after vascular events: a population-based cohort study. *Lancet.* 2017;390:490−499.

26. Scheiman JM, Devereaux PJ, Herlitz J, et al. Prevention of peptic ulcers with esomeprazole in patients at risk of ulcer development treated with low-dose acetylsalicylic acid: a randomised, controlled trial (OBERON). *Heart.* 2011;97:797−802.

27. Bhatt DL, Cryer BL, Contant CF, et al. Clopidogrel with or without omeprazole in coronary artery disease. *N Engl J Med.* 2010;363:1909−1917.

28. Agewall S, Cattaneo M, Collet JP, et al. Expert position paper on the use of proton pump inhibitors in patients with cardiovascular disease and antithrombotic therapy. *Eur Heart J.* June 2013;34:1708−1713.

29. Moayyedi P, Eikelboom JW, Bosch J, et al. Pantoprazole to prevent gastroduodenal events in patients receiving rivaroxaban and/or aspirin in a randomized, double-blind, placebo-controlled trial. *Gastroenterology.* 2019;157:403−412.

30. Lamberts M, Lip GY, Hansen ML, et al. Relation of nonsteroidal anti-inflammatory drugs to serious bleeding and thromboembolism risk in patients with atrial fibrillation receiving antithrombotic therapy: a nationwide cohort study. *Ann Intern Med.* 2014; 161:690−698.

31. Schmidt M, Lamberts M, Olsen AM, et al. Cardiovascular safety of non-aspirin nonsteroidal anti-inflammatory drugs: review and position paper by the working group for Cardiovascular Pharmacotherapy of the European Society of Cardiology. *Eur Heart J.* 2016;37:1015−1023.

32. Storey RF, Gurbel PA, ten Berg J, et al. Pharmacodynamics, pharmacokinetics, and safety of single-dose subcutaneous administration of selatogrel, a novel $P2Y_{12}$ receptor antagonist, in patients with chronic coronary syndromes. *Eur Heart J.* 2020;41: 3132−3140 [Online].

33. Li J, Vootukuri S, Shang Y, et al. RUC-4: a novel $\alpha IIb\beta 3$ antagonist for prehospital therapy of myocardial infarction. *Arterioscler Thromb Vasc Biol.* 2014;34:2321−2329.

34. Quan ML, Pinto DJP, Smallheer JM, et al. Factor XIa inhibitors as new anticoagulants. *J Med Chem.* 2018;61:7425−7447.

A look into the future

Sorin J. Brener, MD

Professor of Medicine, New York Presbyterian-Brooklyn Methodist Hospital, Brooklyn, NY, United States

In the previous 11 chapters, experts from Europe and the United States discussed the development of dual antiplatelet therapy and the role it fulfills in the prevention and management of atherothrombosis. From plaque growth and destabilization to stent thrombosis and myocardial infarction, platelets occupy a critical and central role. Their inhibition is associated with a significant reduction in ischemic events in some vascular beds, while in others the benefit is less impressive or absent. Regardless of the level of protection, increased bleeding always accompanies more powerful platelet inhibition. This "fact of life" is elegantly addressed in Chapter 9. It is also the unresolved piece of the puzzle that surrounds antithrombotic therapy in patients with atherothrombosis and encapsulates the truism that many of the factors that predispose one to ischemic events also enhance the propensity for bleeding. Age and chronic kidney dysfunction are only some of the prominent contributors to both complications of vascular disease and its treatment. In every large randomized clinical trial in the past two decades, more intense antiplatelet therapy was associated with ∼20% reduction in ischemic events and a 30%−50% higher rate of significant bleeding, compared to less intensive therapy. While in general, the number needed to treat was smaller than the number needed to harm, the net benefit was significantly smaller than desired, if at all present.

Thus the future of antithrombotic and antiplatelet therapy must address this problem in one of two ways. Ideally, the antiplatelet therapy of the future will include agent(s) that can separate this inexorable link between ischemic prevention and bleeding causation by identifying specific molecular targets and mechanisms of action. Such hope existed with the introduction of vorapaxar, a protease activated receptor 1 competitive inhibitor, that prevents the activation of platelets by low concentrations of thrombin, without inhibiting hemostasis, in preliminary laboratory and animal testing. Yet, when used in clinical practice, vorapaxar disappointed by causing significant excess bleeding compared to placebo in various clinical settings. Other compounds addressing other potential pathways are far from clinical deployment.

The alternative to this ideal agent(s) is the development of much more precise and clinically useful risk prediction scores. Such scores could potentially identify patients who are at particularly high ischemic risk without prohibitive bleeding risk and recommend aggressive antiplatelet therapy. Conversely, those at low ischemic risk and high bleeding risk would benefit from more conservative therapy. These concepts were well captured in the recent position paper regarding the definition of High Bleeding Risk from the Academic Research Consortium. I am skeptical, though, of the ability of such schemes to delink the connection that exists in many, maybe most, patients between the risk for both ischemic and hemorrhagic complications. A gleam of hope emerged maybe from the Dual Antiplatelet Therapy (DAPT) score in this respect. Beyond its utility in predicting net benefit, the score highlighted an interesting observation. The patients with the highest score (>2) had the greatest ischemic protection (2%–3.5% absolute risk reduction in stent thrombosis and myocardial infarction) with the least excess bleeding (only 0.03% additional GUSTO moderate or severe bleeding) when dual antiplatelet therapy was used for 30 versus 12 months only. The opposite occurred in those with the lowest score (-2 to 0)—minimal ischemic protection (0.07%) and amplified excess bleeding (1.9%). This is the translation in clinical practice of the delinking mentioned earlier. How did that happen? Are platelets of patients with high ischemic risk more resistant to inhibition and thus bleeding does not happen? If so, then, where is the ischemic benefit stemming from? From pleiotropic effects of P2Y12 inhibition unrelated to platelet activation and aggregation? Perhaps, but not proven. Or maybe this is the result of the statistical methods used to build the DAPT score, whereby predictors of both ischemic and bleeding events, such as chronic renal insufficiency or peripheral arterial disease, were removed from the score if they explained too small a portion of the variability observed in net clinical outcome. It is hard to be sure when the score showed only modest accuracy both in the DAPT trial from which it was derived and in the PROTECT study in which it was validated.

What is certain is that we have much more to learn. We invite you to stay tuned.

Index